Playing with Sound

Playing with Sound

A Theory of Interacting with Sound and Music in Video Games

Karen Collins

The MIT Press
Cambridge, Massachusetts
London, England

MIT Press books may be purchased at special quantity discounts for business or sales promotional use. For information, please email special_sales@mitpress.mit.edu or write to Special Sales Department, The MIT Press, 55 Hayward Street, Cambridge, MA 02142.

This book was set in Stone Serif and Stone Sans by Toppan Best-set Premedia Limited, Hong Kong. Printed and bound in the United States of America.

Library of Congress Cataloging-in-Publication Data

Collins, Karen, 1973–
Playing with sound : a theory of interacting with sound and music in video games / Karen Collins.
    pages   cm
Includes bibliographical references and index.
ISBN 978-0-262-01867-8 (hardcover : alk. paper)
1. Interactive multimedia. 2. Video games. I. Title.
QA76.76.I59C653   2013
006.7—dc23
2012025349

10   9   8   7   6   5   4   3   2   1

Contents

Preface: Game Sound Gets Heard

A few years ago, I began my previous book, *Game Sound: An Introduction to the History, Theory, and Practice of Video Game Music and Sound Design* (MIT Press, 2008), by discussing the importance of video games and of game sound in particular. Today, it is taken for granted (at least in some circles) that the study of sound in games is a valid pursuit. Courses are taught on campuses everywhere, and research is attracting wider attention. But in addition to recent changes in the ways in which video games are perceived publicly and in the academy, there have been a variety of changes in the industry, in the technology, within game studies, and in my own personal thinking about game sound that have inspired me to write this follow-up book.

The video game industry has grown tremendously in the past half-decade, and most major game development companies have expanded their sound departments. The roles played by sound professionals have become increasingly specialized as production becomes more complex for "triple A" games—large-scale, epic games with multimillion dollar budgets. But another significant development in the industry is the rise of independent companies, particularly new game companies that have flourished in the social media space. Zynga's *FarmVille* game, for example, managed to attract over 80 million active monthly Facebook users at its peak. Such games have dominated the social networking world and become a ubiquitous form of casual gaming.[1] Multiplayer online gaming has expanded as Internet speeds have improved and social networking and smart phones (notably Apple's iPhone) have gained popularity. Independent game developers from all over the world have found a foothold in the industry by creating game applications ("apps"). The accessibility and affordability of apps has resulted in the growing popularity of casual gaming in general— on smart phones, tablets, netbooks, and other portable devices. To a significant extent, the success of these devices has depended on the availability of free or cheap game apps. Social networking and mobile media have had interesting consequences for game sound, since both forms of casual games are often played with the sound turned off. The eventual outcome of this fact for game sound practitioners remains to be seen, but the history of games is repeating itself once again. Small developers have a space where they can sell and share their games, and the "one-man band" game developer has returned.

Technological changes in the game industry in the past few years have also affected the development of game sound. Audio middleware engines such as Wwise and Fmod enable practitioners to demonstrate the interactive elements of their sound without having to program code.[2] New game sound software such as *PsaiCore* is opening up audio artificial intelligence and computer-aided approaches to composition in games. As artificial intelligence gains ground in the game development community (and as the predictability and planning of narrative and action becomes more difficult), designers are returning to sound synthesis as they explore new methods in procedural generation to be more responsive to the player's actions. Spatial sound has also become more important to console play as home surround-sound systems have become increasingly affordable. And more important than ever before is the increased role that gesture and the player's body play in gaming. With the massive success of the Wii as well as the introduction of Microsoft's Kinect and Sony's Move, gaming is no longer about sitting relatively still in front of a screen but now incorporates whole-body interaction with the game. This gestural involvement is a recurring theme in this book as I explore the implications for the player's embodied experience of game sound.

As the popularity of games continues to grow, scholars have entered game studies from a wide variety of disciplines, which has resulted in interesting interdisciplinary conferences where computer scientists both clash and collaborate with cultural theorists. The study of game sound has seen remarkable growth in just a few years as the literature in this area has increased, the scope of research has widened beyond what was previously available, and new ideas and perspectives have been brought in from different disciplines. With this rise in scholarly interest, game sound design and composition programs and courses have formed in universities and technical colleges around the world, with a need for literature to support these courses.

Finally, my own theoretical perspective has changed during my time at the University of Waterloo, a university known for its strong computer science and engineering programs. I moved out of a music department and into a digital arts program where I now teach games and sound design. Along with my collaborative work with people from the humanities, social sciences, and natural sciences at Waterloo and beyond, this disciplinary shift has altered my own thinking about interactivity and sound through my exploration of less conventional video games (such as slot machines). One of these research collaborations resulted in several trips to Japan, where I wandered the *otaku* alleys of Akihabara and Harajuku and became fascinated with the role that fandom plays in Japanese video game consumption. This helped develop my interest in the player's experience of games.

All of these changes have led me to develop a new approach to thinking about interactive sound that I present here and that in some ways represents a shift from my previous thinking. I still integrate recent research from music, film, and cultural studies, but I expand this disciplinary background to include more work from computer science and psychology. In short, I build on my previous thinking by offering a companion book to *Game Sound*, which focused on the creative, production side of game sound. In this book, I seek to understand games not as texts but as sites of participation and practice where players construct meanings. Here, then, I explore game sound from the other side of the console, focusing on the player's experience of sound.

Acknowledgments

This book originated from a presentation that I made at the Music and the Moving Image conference at New York University in May 2010, where I questioned the lack of research on the player's experience of interacting with sound. The ideas were galvanized after giving another talk a few months later, this time at Concordia University in Montreal, where Chris Salter, author of *Entangled: Technology and the Transformation of Performance* (MIT Press 2010), challenged me with the question, "But what about the body?" I knew then that I was on track to filling a hole in my own previous work.

The bulk of this book was written while on a sabbatical provided by the University of Waterloo, during which time I was a visiting scholar at the University of California at Davis: the support of those institutions is gratefully acknowledged. Faculty at UCD were kind and welcoming, and I especially thank Michael Neff and Colin Milburn for their support, for listening while I worked through some of my ideas, and for thought-provoking conversations during my time there. Some small sections of this book were published in other forms. Parts of chapter 1 are taken from a forthcoming book chapter to be published in the *Oxford Handbook of New Audiovisual Aesthetics*, and parts of chapter 5 are to be published in the *Oxford Handbook of Virtuality*. My master's student Alexander Wharton and I rewrote his thesis paper for *Game Studies*, and elements of that article were incorporated into chapter 5 also. Some paragraphs from chapters 1 and 2 were adapted from a conference paper for *Audio Mostly 2011*.

Colleagues, collaborators, and friends have supported and encouraged me throughout the writing of this book, including arguing with me about some of its points: I am particularly grateful to Bill Kapralos, with whom I regularly collaborate. Also, my gratitude goes to Gillian Anderson, Neil Baker, Brian Cullen, Ruth Dockwray, Lee Ann Fullington, Kevin Harrigan, Ron Sadoff, Stacey Scott, James Semple, Richard Stevens, Philip Tagg, Holly

Tessler, and Mark Wolf. I particularly thank two of the external readers of an earlier version of this book, Mark Grimshaw and Kiri Miller, who both offered insights and suggestions that considerably changed—and strengthened—the manuscript.

I have also been encouraged and supported by many people who work in the industry and were kind enough to answer questions, explore some of my crazier ideas, unwittingly inspire ideas, and provide friendship: thanks to David Battino, Rob Bridgett, D. B. Cooper, Peter Drescher, Gordon Durity, Andy Farnell, Brad Fuller, Chris Grigg, Stephen Harwood, Aaron Higgins, Damian Kastbauer, Michael Kelly, Linda Law, Jennifer Lewis, Rory O'Neill, Leonard Paul, Jory Prum, Jim Rippie, David Roach, Carsten Rojahn, Tom Salta, George Sanger, Brian Schmidt, Michael Sweet, Tom White, Guy Whitmore, and to everyone at Project Bar-B-Q and in the Interactive Audio Special Interest Group (IASIG).

My recent research has received funding from a variety of organizations, and that work has been instrumental in shaping my thoughts about game sound. I therefore offer sincere gratitude to the Canada Foundation for Innovation, Google Inc., Ontario Centres of Excellence, the Ontario Problem Gambling Research Centre, the Ontario Ministry of Research and Innovation, the Ontario Media Development Corporation, the Research Institute of Electronics at Shizuoka University and the Social Sciences and Humanities Research Council of Canada. Thanks especially to my mother for her ongoing support throughout my career and to Douglas Sery, my editor at the MIT Press, for his patience with my revisions and the opportunity to share this work with you all.

Introduction

The first time that I played *Super Mario Bros.* (1985) on the Nintendo Entertainment System was during the Christmas school break of 1985. As a middle-schooler, I marveled that the graphics seemed real to me, but more than that, I enjoyed pressing the buttons on the controller. Instead of actively trying to meet the game's goals as quickly as possible, I made poor mustachioed Mario the plumber leap up and down repeatedly as I reveled in the "bwoop" sound that he made whenever I pressed a button. I had played video games before this one on many different early consoles and in many darkened arcades, but there was something about the sounds made by the Nintendo that endeared the machine to me. I suspect that I am not the only one who feels this way. To this day, Mario continues to make that instantly recognizable "bwoop" as he jumps through countless episodes of games within the franchise. It is not a realistic sound, but sonic reality is not needed in this particular game world. It's not about fidelity: it's about having fun with sound.

This book is about how video game players interact with, through, in, and about sound. It concerns the ways in which meaning is found, embodied, created, evoked, hacked, remixed, negotiated, and renegotiated in the space of interactive sound in games. My aims here are to document these interactions, to discuss the means by which players create their own meanings with and around game sound, to explore game sound from the perspective of the player, and to explore game sound as an experience and practice rather than as a text. This book is about us: the players, the audience, the listeners. It's about the sonic interactions that we have with games and with other players. It's about the complex nature of the interactions between sound, image, and action in the game. It about hacking the machines and the code and making new sonic creations out of them. It's about having fun with sound.

How Is Interacting *with* Sound Different from Listening *to* Sound?

The research presented here stems from a theoretical issue that I raised in my previous book, *Game Sound: An Introduction to the History, Theory, and Practice of Video Game Music and Sound Design* (2008), which is that our understanding of the perceptions of sound in media typically involves a physically passive listener. This is not to say that active listening has not been a component of this research but rather that our physical interaction with sound has been absent in much of the theoretical work about sound. When it comes to interactive media like video games, we lack the terminology and methodologies to study and discuss the players' relationship to sound. In other words, there is much to add to existing theories of sound in media with respect to interactivity. And although academic writing about game sound is growing, we still often miss a fundamental piece of the puzzle that is essential to any theoretical account—the player. Without a player—without the act of play—it is just code, lying in wait. There are good reasons that the player has been absent from many approaches to sound in media: the experiential aspect of interactivity is highly contested. As Lev Manovich (2001, 56) describes, "Although it is relatively easy to specify different interactive structures used in new media objects, it is much more difficult to deal theoretically with users' experiences of these structures. This aspect of interactivity remains one of the most difficult theoretical questions raised by new media."

This book aims to provide a foundation for debate by developing a theory of the interactive sound experience. It is based on my hypothesis that interacting with sound is fundamentally different in terms of our experience from listening without interacting; that there is a distinction between *listening* to sound, *evoking* sounds already made (by pressing a button, for instance), and *creating* sound (making new sounds), a point to which I return later. In particular, I seek to answer these questions: What does it mean to interact with sound? In what ways do game players interact with sound? What makes interactive sound different from noninteractive sound? And how does interactive sound change our association to, our involvement with, and our experience of games?

These questions are explored from a range of disciplinary perspectives, including film studies, game studies, philosophy, musicology, sociology, cultural theory, interaction design, psychology, advertising, neurobiology, acoustics, and areas of computer science. I do not subscribe to any one overarching theory or methodology here. Rather, my ideas are informed by the research in these sometimes disparate areas, by my own empirical studies into interacting with sound, by discussions with those involved in game playing and production, by observation, and by playing a lot of games.

The Sound of Music: Musical Sound

Game *sound* in this book refers to all of the sonic aspects of a game—discrete sound effects, ambient sound beds, dialog, music, and interface sounds. In games, these sonic elements are often closely integrated into the experience of play. Although the relationship between these elements may vary depending on the genre or specific mechanics of a game, generally all of these elements contribute to the overall sonic experience that the player enjoys. Therefore, I believe that theorizing about a single auditory aspect without including the others would be to miss out on an important element of this experience, particularly since there is often considerable overlap between them.

Sound effects may frequently be used in musical composition in games, for instance. This phenomenon is not unique to games and has been discussed at length elsewhere (e.g., Kahn 1999). As in the musical compositions of the twentieth-century avant-garde, the overall sonic texture of games can often create an interesting interplay between music and sound effects that blurs the distinction between the two. In many examples of games, we might question whether we are hearing sound effects, voice, ambience, or music, such as the soundtracks for *Quake* (1996) or *Silent Hill* (1999). Some sound effects in games are also worth considering as music rather than as nonmusical sound. In fact, in many genres of games (and throughout video game history), sound effects have often been tonal (such as Mario's jumping "bwoop" described above). Indeed, sound design in some genres of games is much closer to slapstick comedy and cartoons than it is to most live-action Hollywood films, with music more intimately tied to action, rhythm, and movement and stylized in ways that are often not intended to be realistic. Even in live-action film, famed film sound designer Walter Murch (2005) argues, "Most sound effects . . . fall mid-way [between music and noise]: like 'sound-centaurs,' they are half language, half music." In other words, it is sometimes difficult to make a distinction between music and sound effects in games.

In some games, the sound effects and music are so intimately linked that they are integrated together in the gameplay. In *Mushroom Men* (2008), for instance, the sound effects are quantized to the musical beat, as described by their composer Matt Piersall (in Kastbauer 2011):

Ok, so there's these bees in one of the levels, and we made a track that was the theme for the bees that was like [sings/buzzes theme]. We tied this into the bees particle effect so that its [sic] in sync with the . . . [buzzes like a raver bee]. So yeah, the bees are buzzing on beat, which I thought was really fun. Actually, every ambience in the whole game is rhythmic . . . wood creaks and crickets and all the insects you hear which are making a beat, and every single localised and spatialized emitter based ambient sounds are on beat too.

Although examples in which events are directly quantized to the beat in this way are not common, it is nevertheless not unusual to integrate sound effects and music in a game in some manner.

Furthermore, dialog can also be used in a sonically textural or melodic fashion. Many examples of Nintendo games treat sound effects as voice, for instance, and repeat a pitched sound according to the number of syllables in the accompanying written text. In *Mario and Luigi: Bowser's Inside Story* (2009), low-pitched sounds are associated with large characters (such as the turtle-shelled male character Bowser), high-pitched sounds are associated with small characters (such as the round yellow female Starlow), and Mario and Luigi speak in a gibberish Italian (and the occasional short English phrase). The decisions that the sound (effects) designers, composers, and voice artists make contribute to the ways in which the player experiences the game, and although these contributions are treated as separate entities in this book, when using the term *game sound*, I am referring to the broad collective soundscape of a game.

Interacting with and Listening To

What does it mean to interact with sound? I begin with a simple semantic distinction: we listen *to*; we interact *with*. The assumption in this use of terminology is that when we interact *with*, we are a participant in an action, whereas when we listen *to,* we are external to the action taking place: we are an auditory observer. But does interacting with sound change our subjective experience of that sound? And if so, how does this altered perception affect the player's game experience?

Our perception of sound is affected by how we listen to that sound. Listening is not merely the act of hearing sound but also is consciously attending to sound. In relation to film, theorist Michel Chion's categorization of three basic listening modes is often cited as a means to understand the act of listening. Chion (1994, 28) describes first the most common mode of listening, causal listening. Causal listening refers to the act of focusing on or recognizing the cause or source of the sound. We constantly, consciously or not, gather information based on the sounds that we hear—where a sound is located in our environment, what type of object caused the sound, the general characteristics of that object, and so on. In comparison, semantic listening refers to the ability to listen to and interpret a message in the sound that is bound by semantics, such as spoken words. Causal listening and semantic listening are not mutually exclusive; we listen to both the words that someone says as well as how someone says them, for example. Finally, reduced listening denotes listening to the traits of the sound (that is, the acoustic properties such as quality or timbre of a sound), independent of cause and meaning and independent of its environmental context. For example, in the game *Fallout 3*, a broadcast tower sends out beeping signals: if I listen causally, I likely assume that some form of electronic equipment is making the sound in the game, off to the right, at a perceptual distance of thirty meters. If I listen semantically (and understand Morse code), I may listen to the message. And if I listen in a reduced fashion, I may describe the sound as a smooth sine wave at upper midrange frequency in short bursts of approximately one half-second each. Listening affects the ways in which the player experiences the game and, in some cases, affects the player's ability to play the game. As described above, however, these modes of listening are not mutually exclusive, and a player may be listening in several ways at any one time while playing a game.

Nevertheless, these three listening modes fail to capture many of the other ways in which we may also attend to sound, as Chion himself acknowledges. Notably, when it comes to interactive media, some ways of listening imply action and participation on the part of the listener. Musicologist David Huron (2002) has produced a more exhaustive and detailed list of listening modes that apply specifically to music. A few noteworthy additions to Chion's list include signal listening, sing-along listening, and retentive listening. Signal listening might also be referred to as anticipatory listening or, according to Huron, "listening in readiness." With signal listening, we listen for a particular cue or auditory sign post: the player is anticipating a signal to action. For example, in the video game *New Super Mario Bros.* (2006), the player sometimes must listen to the music to time Mario's attack, since enemy Goombas jump in time to the beat. We also can extend signal listening beyond music by listening to sound effects for navigational information about direction, proximity, and spatial cues; status information about a process or event; and semiotic information on the nature of virtual characters that we meet or places that we encounter.

Sing-along listening involves mentally or physically following along with sound. Popular karaoke games like *SingStar* (2004), in which the player sings into a microphone with a song and is judged on accuracy of timing and pitch, is a case where players employ sing-along listening. More generally, for example, we may harmonize to a song that is playing while driving in a car. As I discuss further in chapter 2, sing-along listening is not necessarily singing out loud. Mental mimicry of singing along can also be categorized as a form of sing-along listening.

Retentive listening occurs when we try to remember what is being heard, usually for the purpose of repeating it. The *Simon* (1978) game asks players to remember simple sequences of tones to be prepared to repeat them, and so they listen with the intent of retaining the pattern of tones. Depending on the type of game, players may listen to those sounds causally (with the intent of remembering a sequence of animal sounds, for example), semantically (with the intent of repeating a particular pattern of meaningful information, for example), or in reduced fashion (with the intent of identifying or repeating particular elements or frequencies of a sound). In other words, Huron's listening modes may be umbrella modes that encompass Chion's methods of listening. All of these additional listening modes suggest that some ways of listening to sound anticipate an action that we are going to take. The sound implies a subsequent activity on the part of the listener (playing the next part, singing along, repeating a sequence): in other words, these are all interactive ways of listening to sound.

The majority of our interactions with sound occur when we are not just listening and reacting to sound but also when we are evoking, selecting, shaping, or creating a sound. In evoking sound, we call forth a preexisting sound object (for example, by triggering a musical sequence or a sound effect in a game). If I press a button and hear Mario jump, I am evoking the Mario jumping sound. Not all sound evoking may be as self-evident as this example, however. I might set in motion the real-time generation of a sound that incorporates game parameters over which I may have some control (shaping the sound), or I might do something that causes a particular sound to be played (intentionally or not). In this instance, I am moving toward creating a sound—that is, generating a sound without a predetermined context within which the sound is to be heard. For example, when I improvise (create) sounds on a musical instrument, the sounds are freely generated and not organized according to a context. The distinction is subtle, and the difference can be thought of as existing on a continuum from evoking (least creativity or freedom) to creating (most creativity and freedom.

My contention is that the experience of interacting *with* sound is fundamentally different in terms of the listener/player's experience from that of listening *to* (noninteractive) sound. But to explore this hypothesis further, we must undertake an in-depth exploration of the types of interactivity that occur between a player and the sounds of a game in terms of the player's perceptual experience.

Definitions of interactivity have been debated for several decades now, particularly with the increased interest in game studies that has developed in recent years. But even earlier, with the rise of audience-centered cultural studies in the 1980s, the concept of interactivity took on new meaning as researchers argued that the process of making meaning from texts is a form of interactivity and that this psychological interaction is as valid as any form of physical interaction. Indeed, it has been argued by some that what was previously called *active reception* is the same as *interaction* and that by its definition "interactivity involves decision making or the active participation of a user" (Morse 2003, 16–17). Others argue that "this interaction remains a mental event in the viewer's mind when it comes to experiencing traditional art forms: the physicality of the [text] does not change in front of his or her eyes" (Paul 2003, 67).

Despite the debates on the definition of interactivity, many theorists place agency and the ability of the media to respond physically to the audience's action as central elements of interactivity. In describing musical instruments, it has been suggested that "interaction between a human and a system is a two way process: control and feedback. The interaction takes place through an interface (or instrument), which translates real-world actions into signals in the virtual domain of the system. . . . The system is controlled by the user, and the system gives feedback to help the user to articulate the control, or feed-forward to actively guide the user" (Bongers 2000, 128). We might reason that this control and feedback/feed-forward are at the heart of an interactive experience and therefore designate active reception as an active but not *inter*active experience.

But interactivity is not necessarily a better experience than those that we have with noninteractive media. In their book *Digital Play: The Interaction of Technology, Culture, and Marketing*, Stephen Kline, Nick Dyer-Witheford, and Greig de Peuter (2003, 14) make the case that interactivity has been "represented as a dramatic emancipatory improvement over traditional one-way mass media such as television and its so-called 'passive' audiences." They also acknowledge, however, that with video games, "clearly there is an important cultural shift taking place from spectators to players" (18). In other words, the experience of physically interacting with media is somehow different, if not necessarily superior.

One difficulty with defining interactivity is that a single media object or text may be fluid in its degrees of interactivity and may afford different degrees of interactivity at different times. This fluidity suggests that there are a variety of different types of interactivity that take place with media such as video games. Even when discussing interactive media such as games, there are different degrees of involvement and interactivity, and it is not the fact of interaction but the ways in which we interact that is most important to our conceptions of interactivity. Therefore, it may be helpful to consider interactivity on a nonhierarchical spectrum as shown in figure I.1. Some of these interactions take place directly between a single player and the game (in-game), and some are largely external to the game play (the metagame interactions).

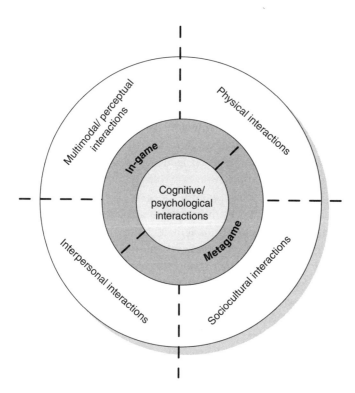

Figure I.1
Sonic interactivity in a game.

Although cognitive/psychological interactions do not directly involve physical control or feedback, such interactions always occur alongside all other types of interactions in games. These interactions include the conscious or unconscious emotional and cognitive activities that take place before, during, and after gameplay. Psychological interactions, therefore, can be thought of as the center of a circle from which all other interactions take place. As noted above, some theorists posit active reception (a cognitive/psychological activity) as a form of interactivity. Manovich (2001, 57) reasons, "When we use the concept of 'interactive media' exclusively in relation to computer-based media, there is the danger that we will interpret 'interaction' literally, equating it with physical interaction between a user and a media object (pressing a button, choosing a link, moving the body), at the expense of psychological interaction." However, Manovich falls into the Cartesian trap of assuming a separation of the mind from the body, since psychological interaction exists only with the physical interaction of the body (Murphy 2004, 228). As is shown in this book, an embodied interaction with games is key to understanding our psychological interactions with them also. Brain and computer interactions and psychophysical sensors like the Wii Vitality wand suggest that mentally or psychophysiologically controlling games is now possible, and experimental games like *Alpha World of Warcraft* (Plass-Oude Bos et al. 2010) have demonstrated that players can use their alpha brain waves to change gameplay, thus further confusing any division between the physical and psychological. In these instances, is the mind or the body controlling the game? I include cognitive and psychological interactivity here, therefore, because it is always an element of all other forms of interactivity.

Beyond psychological interactions, other types of interactions have been divided tangentially into interactions between the player and the game (perceptual or multimodal and physical interactions, defined below) and metagame interactions (interpersonal and sociocultural) that take place around and extend beyond the game. Although separated here for discussion purposes, these categories are not mutually exclusive and in many cases are interdependent.

As Manovich described above, some definitions of *interactivity* tend to refer narrowly and mechanically to what might be called *physical interactivity*. David Z. Saltz (1997, 118), for example, suggests that three events must occur for interactivity to exist: an input device (or control device, such as a joystick) must capture a person's behavior, the computer must interpret the input, and an output device must translate the data back to the person (via a feedback device, such as the gameplay). Relying on an equally mechanical definition, Matthew Lombard and Jennifer Snyder-Duch (2001) identify five variables that influence the degrees of interactivity, and all depend on direct physical manipulation, control, and feedback—number of inputs, number and type of characteristics that can be modified, extent of change possible, speed with which the medium responds to the user input, and degree of correspondence between input and response. Although physical interactivity is perhaps the most obvious form of interactivity and is considered by many people to be the only form of interactivity, in this book I take a much broader approach to try to understand all of the ways in which we interact with sound. For instance, there are also interactions that take place between modalities (audio, visual, and haptic). The player in this case is involved in the perceptual process of interpreting these interactions, both as a spectator and as a causal agent (interactor).

In addition to the in-game interactivity that takes place, interpersonal interactions take place in—and are mediated by—the game. Interpersonal interactions occur between players as character avatars and between the players themselves. Interpersonal interactions therefore take place both internally and externally to the game. Players in multiplayer games might know each other outside the game world and use the game as another mode of communication, and some activities that take place in and around games may not be directly related to the game, such as the social talk that occurs between players (Taylor 2006, 84–86). Much evidence suggests that players enjoy games for social reasons. For example, local area network (LAN) parties (social gatherings where groups of players play a game together over a local area network) still exist even when it is possible to stay home and play the same game online, suggesting that players enjoy these colocated interpersonal interactions.

Interpersonal interactions may also be extended into larger sociocultural interactions. These may happen on a much wider and temporally longer scale, including interactions between the game designer and the player, for instance. These interactions may include criticism of and commentary about the sound on blogs, magazines, and so on. The designer may respond to player criticism and provide patches, updates, or changes in a sequel, thus interacting with players in a less direct fashion. For example, a preview beta version of *Assassin's Creed* was demonstrated to a live audience at the 2007 Electronic Entertainment Expo (E3) (and subsequently broadcast online). Player reaction to the game preview was disappointing, particularly in reference to what was perceived as poor collision detection.[1] The developers quickly engaged in a public relations campaign and declared that the problem was fixed in the debugging cycle and that the subsequent release of the game responded to players' criticisms (Newman 2008, 40–41). This example illustrates how player feedback during a prerelease preview can influence a game. Developers have also actively sought out player feedback. With Sony Online Entertainment's Guild Summits for *EverQuest* (1999), players were brought in during the development cycle of extensions to the online game to comment and provide input. Players who could not attend the Guild Summits published online open letters to have their criticisms heard (see, for instance, Wolfshead 2004). Significant changes were made to the game as feedback was incorporated.

Players may also use video games for nongame purposes or customize games in ways that were either intended or unintended by the designers. All of these interactions among player, game, and designers may then become part of a larger cultural dialog about music, games, art, politics, and so on. These culturally productive forms of interactivity situate the interactive audience as somewhere between the author/producer and the audience/consumer, and this relationship is played out in courts of law as much as in cultural theory when developers feel that they have lost control of their intellectual property.

The new and altered creative materials that develop from games have been a significant focus of game scholars. The notion of players as fans has brought from media studies the idea that players may create their own meanings from texts but that fandom involves textual productivity (Fiske 1992). This conceptualization of fans as necessarily productive is problematic, however, particularly due to the narrow notions of productivity that are involved. Hanna Wirman (2009), for instance, criticizes the notion of play as unproductive, suggesting that the idea is perhaps a holdover from early play studies. She argues that if players can configure, explore, and add content to games, they become cocreators of the game and producers of meanings, providing "partial authority" to the player. The result of this (inter)activity means that "games are better understood as platforms for experiences than as products, but also that games as cybertexts are only partly predetermined or precoded before the activity of play takes place" (Wirman 2009). If our understanding of productivity is widened, then all players are productive and therefore fans, according to Fiske's notions of fandom.

Nick Couldry (2004) argues that a new shift in cultural theory was anticipated by the idea of mediation, in which theorists study social processes and practices that relate to media. Mediation can be defined as an understanding of what media do and what we do with media (Silverstone

2003). Thus, the audience becomes an (inter)active component of meaning making in the consumption of any media text. Other similarly audience-driven theories have arisen in the past decade. Practice theory views media as a set of practices and discourses as a component of social life and asks, "what types of things do people do and say in relation to media?" (Couldry 2004, 121). Practice theory removes media from the study of production and industrial structures and shifts focus to the audience's productive and consumptive practices around media. Instead of emphasizing the text or the audience, the focus on practice can "help us address how media are embedded in the interlocking fabric of social and cultural life. . . . Through this, we can perhaps hope to develop a different approach towards understanding media's consequences for the distribution of social power" (129). Practice theory expands beyond communication theory into the sociology of action and knowledge and into cultural and cognitive anthropology (Couldry 2004).

The change in theoretical approach in cultural studies over the past few decades has been influenced by a cultural shift toward more interactive media. Interactivity by its nature complicates the author/audience divide and confuses the notion of the text as a finished product (Saltz 1997, 117). Interactive texts are inherently unfinished because they require a participant with whom to interact before they can be realized in their myriad forms: the player is needed for the game. The structures that are inherent in interactive media encourage both a greater capacity and a greater interest on the part of the audience toward coauthorship through altering and manipulating texts. Since interactivity has the capacity for the audience to be involved in, engage with, shape, or customize the text, it thus "spurs on and sometimes encourages a desire to transform the text in ways that are out of the hands of an author and in accord with the individual wishes of an audience member or user" (Cover 2006, 141). Interactivity therefore causes tensions between author and audience that give rise to attempts to control authenticity and authorial "purity" through intellectual property rights or digital rights management. Digital interactive media thus actively encourages attempts to circumvent controls to facilitate that customization and personalization of the text by the audience (Cover 2006).

Marshall McLuhan and Barrington Nevitt predicted as early as 1972 that the consumer/producer dichotomy would blur with new technologies. Futurologist Alvin Toffler (1970) coined the term *prosumer* (producer consumer) to describe this blurring of roles. Today, the term *cocreative media* (Morris 2003) acknowledges the fact that neither the developers nor the game players are solely responsible for the production of "the game"—in other words, that games require the input of both players and developers. In situating the ascendency of the prosumer, Rob Cover (2006, 146) argues that "the rise of media technologies which not only avail themselves to certain forms of interactivity with the text, but also to the ways in which the pleasure of engagement with the text is sold under the signifier of interactivity is that which puts into question the functionality of authorship and opens the possibility for a variety of mediums no longer predicated on the name of the author." Interactive media technologies arguably represent the democratization of control over the text. But Cover goes too far in suggesting that the "conception of the author-text-audience affinity can be characterized as a tactical war of contention for control over the text" (140–141). The relationship between original creator and the new cocreator is mercurial, and game companies are gradually providing affordances for—and, indeed, encouraging—customization, modification, and other forms of play with the game.

Here, I combine recent ideas and theories about media audiences to examine and describe game sound as a site of practice and productivity in which sound becomes method, material, and mediator of experience. I argue that the interaction with game sound is in a unique position in terms of media audiences, where multimodality, spatiality, and social interaction intertwine to create a plurality of meanings. Taking the torch from mediation and practice theory, I explore the extension of game sound beyond the game as text to the interpersonal and sociocultural interactions that occur in and around games. I examine the practices that surround the audience's construction and manipulation of sound in games; the creation of music in games, music from games, and music about games; the creation of other media from game sound; and the creation of games from sound. In this way, sound becomes the medium through which interpersonal and sociocultural interactions take place.

Game Players: An Interactive Audience

Game players are an active audience and enjoy sound not through passive listening but through physical and social interactions. But game players are as different as the population at large. A misconception often perpetuated by the media is that gamers are mostly teenaged males who play alone. Recent statistics have shown otherwise. According to the industry's dominant spokesgroup, the Entertainment Software Association, a higher percentage of females over the age of eighteen play games than males under the age of eighteen.[2] Although home console (Xbox, PlayStation, Wii) game play still tends to be somewhat male-dominated, the game audience as a whole is nearly evenly split between men and women. And rather than a solitary pursuit, today games are primarily a social experience and played in groups or pairs (virtually via online gaming or in the same physical space). Some evidence suggests that as players become more involved in an online massively multiplayer game, they do not become more socially isolated (as the popular press might have us believe) but instead become more involved with the social networks that surround the game, using the Internet to discuss and share ideas and knowledge about the game and to plan collective play (Taylor 2006, 81).

As a result of the different markets within the game industry, some theorists, game companies, and industry organizations have developed designations of players that span gender and age distinctions. These began with simplistic dichotomies of novice and expert and of casual and hardcore (sometimes with a moderate level in between), where casual gamers are occasional players and hardcore gamers purchase and play console games extensively (Sotamaa 2007). Marketing firm Parks Associates (2006) developed six designations: power gamers regularly purchase games and account for 30 percent of sales, social gamers play with friends, leisure gamers are largely casual gamers, dormant gamers like to play but do not have the time, incidental gamers may play casual games but do not purchase games, and occasional gamers pick up a game once in a while. The difficulty with this type of purchasing-based division is that players may move between these groups depending on changes in their specific social and financial situation regardless of their desire to play, and it remains unclear if a meaningful distinction can be made among leisure, incidental, and occasional gamers.

Rather than focus on how often people play or purchase games, other theorists have divided players into the types of play in which they engage. Chris Bateman and Richard Boon (2006) divide players into conquerors (who focus on winning), managers (who enjoy strategy), wanderers (who play for fun), and participants (who play for social or narrative reasons). Katie Salen and Eric Zimmerman (2003) divide players into groups based on the player's relationship to the rules of the game: dedicated gamers develop their own unique strategies, unsportsmanlike players have bad attitudes, standard players follow the rules, cheats violate the rules, and spoil-sports try to ruin the game. Perhaps the best-known categorization of players is that of Richard Bartle (1996), who focuses on the psychology of players, grouping them according to socializers (who use the game to communicate and socialize), explorers (who like to map out and know the game world completely), killers (who distress other players), and achievers (who adopt game-related goals and set out to achieve them). Bartle (2004) later adds eight further categories of motivations for play. Most useful, however, is Bartle's assertion that players shift between these groups regularly—even within a single play session—with the implicit assumption that not everyone plays games for the same reasons and that some players play in more than one manner. As is shown in later chapters, however, many forms of play occur outside the direct interaction between player and game, and these forms of play must also be considered in any discussion of player type.

Game players are not an amorphous mass but are individuals whose involvement, motivations, and playing style may be constantly in flux. My interest here is in exploring the many ways in which players can interact in, with, and around game sound rather than in defining who is engaging in these types of interactions. For my purposes, then, the game player is anyone who interacts and plays with and around games.

Game Players: An Embodied Cognition Approach to Audience

Because interactivity is both a physical and psychological engagement with media, we should approach games from theoretical perspectives that take into account the physical and psychological aspects of experience. There are many theories about our perception of everyday phenomena. For many centuries, the dualism theory of René Descartes dominated Western thought, and it was believed that the mind and body were separate: the mind was immaterial and thus distinct from the material body. In *Meditation VI*, Descartes (1641/1998, 54) described, "I have a clear and distinct idea of myself as a thinking, non-extended thing, and a clear and distinct idea of body as an extended and non-thinking thing." Into the early twentieth century, Western thought continued to support the Cartesian dualist point of view. Early theories of cognitive psychology separated the mind from its material existence: cognition was "embrained" in the physical brain but not in the body (Damasio 2000, 118). Classical cognitive theories allowed "sensory, motor, and emotional experience to be represented as stripped of their perceptual and experiential basis" (Niedenthal 2007, 1002). One such theory, computationalism, held that the mind was similar to a computer. The mind, like a computer's central processor, receives sensory data from the body's input devices (sensory perception) and then manipulates that data to form ideas. In this way, our perception of the world is indirect, since we understand the world only through our "data analysis."[3]

Philosopher Edmund Husserl (1931/1962) largely discarded the cognitivist viewpoint and took a step toward what came to be known as *phenomenology*, defined as "a philosophy concerned with the interpretation of human experience rooted in perception and bodily activity" (Ihde 1986, 60). For Husserl, consciousness is based on the phenomena that arise from physical, corporeal experiences. The mind, therefore, is shaped by our physical existence and experience of the external world, although Husserl still separated mind from body. Martin Heidegger (1927) and Maurice Merleau-Ponty (1945) further developed the phenomenological approach, rejecting Husserl's Cartesian influence and instead focusing on the role that the body plays in our consciousness. For Heidegger (1927), this was predicated on our being embedded in our environment ("being-in-the-world"). Merleau-Ponty (1945) similarly held that all perception is understood through the ways in which we are able to act in the world and move around in our environment (the body as lived or "lived-body," in the world as lived or *lebenswelt*). In this way, the sensory input that we receive from the environment forms our understanding of the world, and our body acts as a mediator between the world and our consciousness.

After a few decades of falling somewhat out of philosophical favor, phenomenology has recently seen a resurgence in popularity, particularly as cultural theorists try to make sense of our embodied experience of interactive media. Don Ihde's *Bodies in Technology* (2002), for instance, proposes three ways of understanding embodiment. Ihde makes a distinction between the "sensory body" (the physical, material body of spatial orientation, physical perception, and emotion) and the "cultural body" (the socially, politically, and culturally constructed body). In a way, this theory is an extension of Husserl and Merleau-Ponty's division of the body into the physical material body and the lived experience. But Ihde also describes a third body that combines the first two and situates this in the environment: the "technological body" is interactive with technology, whether with primitive tools or high-tech devices. In this way, the body can be extended through the use of tools. Ihde (1986, 141) calls this "extended embodiment"—an "instrument-mediated experience in which the instrument is taken into one's experience of bodily engaging the world." Thus, our sensory perception is extended through our technologically mediated experience of the world.

A new perspective that combines the philosophical phenomenological experience with scientific cognitive theories of perception has begun to emerge over the last decade or so. Arising out of a concern for understanding the mind in its corporeal context, embodied cognition holds that our understanding of the world is shaped by our ability to interact with it physically. Embodied cognition is "the creation, manipulation and sharing of meaning through engaged interaction with artifacts" (Dourish 2001, 126). In this way, embodied cognition is complementary to phenomenology, but whereas phenomenology is based in philosophy, embodied cognition is grounded in psychological theory and cognitive science. Mark Leman (2008, 14), who has explored an embodied cognition approach to music, emphasizes the importance of the body in the embodied cognitive approach: "The concept of action allows sufficient room for taking into account subjective-experience and cultural contextualization, as well as biological and physical processes. Actions indeed are subjective: they can be learned, they often have a cultural signification, and they are based on the biomechanics of the human body. In that sense, actions may form a link between the mental and physical worlds."

In embodied cognition theory, our cognitive processes use reactivations of sensory and motor states from our past experience. In other words, our knowledge is tied to the original neural state that occurred in our brain when we initially acquired that information. Our brain captures modality-specific states during the initial perceptions and actions we take and then recalls and reinstantiates those states when needed (Niedenthal 2007, 1003). Our cognition is therefore embodied because it is inextricably bound to our sensorimotor experience, and our perception is always tied to a mental reenactment of our physical, embodied knowledge. Recent empirical research in developmental psychology seems to support the embodied cognitivist view. Our direct physical interactions with the world through our sensorimotor capacities are key to shaping our cognitive development in childhood, for instance (Klemmer, Hartmann, and Takayama 2006). But embodied cognition is a contentious theory that is still in its early stages with guiding principles that are in flux, especially as applied to sound.

The chapters that follow combine an embodied cognition approach with practice theory to explore the interaction and perception of sound in games. In chapter 1, I investigate the in-game interactions between modalities, exploring interactivity's effects on the player's multimodal experience of sound, image, and haptics. Chapter 2 takes this a step further by exploring the role of interactive sound in creating an emotional experience through immersion and identification with the game character. Chapter 3 continues to explore the role that embodied interaction with sound plays in identification and immersion in the game, focusing on the ways in which sound acts as a mediator for a variety of performative activities. In chapters 4 and 5, I expand the discussion of embodied interactions with sound beyond the game, examining the effect that cocreative, performative practice with game sound has on immersion and meaning generation in games.

1   Interacting with Sound: A Theory of Action, Image, and Sound

"Interactivity" is one of those neologisms that Mr Humphrys likes to dangle between a pair of verbal tweezers, but the reason we suddenly need such a word is that during this century we have for the first time been dominated by non-interactive forms of entertainment: cinema, radio, recorded music and television. Before they came along all entertainment was interactive: theater, music, sport—the performers and audience were there together, and even a respectfully silent audience exerted a powerful shaping presence on the unfolding of whatever drama they were there for. We didn't need a special word for interactivity in the same way that we don't (yet) need a special word for people with only one head. (Adams 1999)

In his article "How to Stop Worrying and Learn to Love the Internet," science fiction author Douglas Adams (1999) suggests that our understanding of new media and interactivity is flawed because interactivity is anything but new and is now relevant only because we temporarily turned away from it in favor of more passive entertainment. But what is new about interactivity now is the technology that mediates that experience and the ability of that technology to fundamentally alter the ways in which we interact with media. For instance, we have the ability to separate sound from its source—what Canadian soundscape composer R. Murray Schafer (1969, 43–47) refers to as a *schizophonic* activity. Prior to the advent of recording technologies, sounds were tied to the physical objects or mechanisms that produced them, but the introduction of recording devices enabled the separation of sound from its source—in other words, the separation of sound from the image and gesture that are associated with the causality of that sound. Schafer's terminology comes from the concatenation of the terms *schizo* (split) and *phonic* (sound), implying an anxiety or a pathological nature of the split through its terminological similarity to *schizophrenia*. The sounds that result from the separation of sound from source, in the ears of Schafer, are unnatural and alien: they are disembodied. In earlier cases, when sounds could be disassociated from an emitter (for instance, voices in a highly reverberant space like caves and cathedrals), these sounds were often associated with otherworldliness and unnaturalness. This sentiment is echoed by Gregory Whitehead (1991, 85) in his description of radio, a technology that splits sound from its visual source: "The circularity of cutting into/casting out radiobodies gives radio performance an inescapable post-mortem quality; man is sick because he is poorly made. Each radio transmission embraces the post-mortem recollection of beings that have been physically dispersed across multiple generations of media abstraction."

However, as described by Paul Sanden (2009, 17), "such accounts of recordings as agents of disembodiment promote an overriding technophobia that ignores the very real potential for sound technology to further increase a listener's engagement with corporeality. . . . To assume that sound technologies interrupt the corporeal significance of sounds simply because they remove these sounds from their visual sources is thus to ignore corporeally sensitive techniques of listening that have little if anything to do with sight." We don't just see sounds occurring: we feel them. Moreover, such disembodiment is now sometimes preferred in contemporary popular music—what has been referred to as a component of "fetishistic audiophilia" in studio production (Corbett 1990). Through various studio techniques, the performer's body is removed from the production and eliminated from the listeners' mental image of a performance. John Corbett (1990, 84) describes an "audio-visual disjunction" that results when listeners attempt to fill in the visual gap—"the menacing void"—in music recordings through the production of visual accompanying material, such as music video and album covers.

More remarkable than the separation of sound from source, however, is our ability to reassociate or integrate that schizophonic sound with a new visual source to create new meanings, what film sound theorist Michel Chion (1994, 63) refers to as *synchresis*—"the spontaneous and irresistible mental fusion, completely free of any logic, that happens between a sound and a visual when these occur at exactly the same time." In this way, the fusion of sound and image leads to new meanings that may alter or add to the original meanings of the sound and the image. This separation and subsequent integration of sound and its causal agent are central to most sound design in today's audiovisual media, whether live-action film, animation, or video game. For instance, we hear a recording of a stalk of celery being snapped, but through its association with a visual image of a bone breaking, we hear that sound as a bone break. Free of the ethereal quality that allegedly plagues schizophonic sound, synchretic sound images are given a new form and thus a new embodiedness.

Past research into the relationship of sound to image in cinema has, for the most part, ignored our bodily associations with the media. Recent theory has sought to correct this disembodiment by employing a multisensory approach to the experience, as Vivian Sobchack (2004, 67–68) describes in *Carnal Thoughts: Embodiment and Moving Image Culture*:

First of all, in the theater (as elsewhere) my lived body sits in readiness as both a sensual and sense-making potentiality. Focused on the screen, my "postural schema" or intentional comportment takes its shape in mimetic sympathy with (or shrinking recoil from) what I see and hear. If I am engaged by what I see, my intentionality streams toward the world onscreen, marking itself not merely in my conscious attention but always also in my bodily tension: the sometimes flagrant, sometimes subtle, but always dynamic investment, inclination, and arrangement of my material being. However, insofar as I cannot literally touch, smell, or taste the particular figure on the screen that solicits my sensual desire, my body's intentional trajectory, seeking a sensible object to fulfill this sensual solicitation, will reverse its direction to locate its partially frustrated sensual grasp on something more literally accessible. That more literally accessible sensual object is my own subjectively felt lived body.

Sobchack defines what she terms a "cinesthetic" experience of cinema, a play on the neurological condition synesthesia, in which sensory inputs are confused and integrated. But despite this turn toward a corporeality of the cinema, most of the work relies nearly completely on the visual-corporeal connection of the audience to the film. Indeed, the titles of many of these works—*The Cinematic Body on the Visceral Event of Film Viewing* (Shaviro 1993), *The Tactile Eye: Touch and the Cinematic Experience* (Barker 2009), and *The Address of the Eye: A Phenomenology of Film Experience* (Sobchack 1992)—belie the authors' allegiance to the ocular. In *The Skin of the Film: Intercultural Cinema, Embodiment, and the Senses*, Laura Marks (1999) offers a theory of "haptic visuality" in which the visual functions like touch, evoking memories of smell, touch, and taste, but she has little to say about sound. Jennifer Barker (2009, 81) at least acknowledges some of the role of sound when she writes, "Often, a film encourages a muscular gesture in the viewer and then expresses its empathy with us by performing the same gesture itself. This happens whenever a sound occurs off screen, cueing us to look (or, to want to look) in the appropriate direction of the image, only to have the film cut to a shot of the source of the sound." She later suggests, however, that "it is the combination of our gaze at the movie screen and our muscular body's commitment to the film space that allows for our feeling of being there, and it is by looking away from it that we break the connection" (86), completely ignoring the fact that sound continues to exist in our space after the visual connection is broken. Reading these works, one might assume that cinema is only a visual media form, existing in near silence. We are still confined, academically, to *watching* a film.

Such ocularcentrism has likewise plagued video game studies, so much so that Bryan Behrenshausen (2007, 335–336) argues that "the discipline of game studies, too, exhibits a near-exclusive preoccupation with video games' relation to players' embodied sense of sight at the expense of exploring other powerfully carnal modes of player–game engagement." The game-playing experience of embodiment is significantly different from that of viewers' corporeal relationship to film. In film, the audiovisual may make filmgoers believe a certain way about objects, characters, or situations (they may hear a bark and think, "That is a friendly wolf" or "That is an aggressive wolf"), but in games, sound makes players behave in a certain way. Their belief is carried a step further into action because they must rely on the audiovisual to navigate a game successfully. Put in simple terms, the stakes for players' involvement, interpretation, and therefore attention are much higher in games, so they listen more actively and employ different modes of listening to guide their own movements and actions in the game. Although film may act *on* the body, players act *with* games, and thus the physical connection with games is distinct and fosters a two-way interaction. Moreover, the added haptic involvement, physical interface devices, and extended auditory environment (with the Wii remote speaker, for instance) create additional multimodal interactions between vision, audio, and haptics.[1] But how is this multimodal experience altered by interactivity?

Sound in interactive media such as games is multimodal—that is, it involves the interaction of more than one sensory modality and usually contains three (vision, audition, and haptics—action, image, and sound).[2] In fact, we could even go so far as to say that, unlike noninteractive sound, interactive sound requires more than one modality. Nevertheless, as I argue below, bringing the interactivity of the player into theories of separation and integration of sonic sources requires a new terminology—and indeed, a new theory—to account for the role of the body and its interactivity with sound. In this chapter, I explore game sound as a component of an interactive multimodal experience that includes haptics and visuals. I examine the effect that sound has on these other sensory modalities and how these other modalities affect players' experience of game sound. To return to the original question ("How is interacting *with* sound different from listening *to* sound"?), my focus is on the role that interactivity has played with the multimodal experience of gameplay.

Schizophonia: Disembodied Sound?

Schafer's (1969, 91) use of the term *schizophonia* suggests that when we hear a sound separated from its source, we might experience some technological anxiety, a fear of "machine-made substitutes" for "natural sounds." But as I show throughout this and the following chapters, sound is never fully separated from its source in our minds. We can never *not* listen causally to sound: it is the dominant mode of hearing. In other words, even when presented in a single modality, sound on its own is, in a way, multimodal. For instance, Mark Leman (2008, 139) contends that music involves all of the senses and that it "moves the body, evokes emotional responses, and generates associations with spaces and textures." My own prior exploration of this phenomenon, for instance, illustrated that percussive sound effects are semiotically loaded with references to other modalities. When participants in an experiment were asked to freely associate with the percussive sound effects that were played, many reported that the sounds connoted the movement (imagined cause) of the sound (such as a person striking an anvil) or connoted a texture of the sound (rough, cold, and so on) (Collins 2002, 395–396). Experiments by Trevor J. Cox (2008) suggest that the sounds that we dislike the most could be due to our haptic associations with those sounds: we dislike the sound of scraping fingers on a blackboard because we sense how it feels to do this as we hear the sound. In other words, sounds on their own can evoke images and have corporeal associations with their causality.

Indeed, many people listen to radio drama because, like reading a book, the imagery for the story unfolds in the mind (Cazeaux 2005, 157). Clive Cazeaux argues with German film critic Rudolf Arnheim, who stated in the 1930s that radio "seems much more sensorily defective and incomplete than the other arts—because it excludes the most important sense, that of sight" (in Cazeaux 2005, 158).[3] Cazeaux draws on Merleau-Ponty to argue that our five senses are not five discrete channels to input data to the mind but, rather, that the senses operate in a synesthetic unity: we listen and simultaneously see and feel the drama. And although we always listen causally, we are simultaneously interpreting and associating sounds with other events, objects, and emotions from our experiences. For example, we may hear footsteps getting louder, but those sounds both serve as an index for the event (someone is approaching) and also represent to us the wider associations that are related to them (such as gender, authority, or threat) (Cazeaux 2005). Together the phenomenological and semiotic approaches seem to indicate that sound's connotative abilities are due to the synesthetic, sensory integration that takes place in the mind and also to the personal and cultural associations that we have with those sounds.

Although we have an embodied response to radio drama through the synesthetic effects of sound, most radio dramas lack any direct physical interactivity. So how is interacting *with* sound different from listening *to* sound? As a point of comparison, consider audio-based games, which are games that do not rely on visuals. They generally can be played without graphics by the visually impaired. The idea of audio-based games goes back to at least *Real Sound—Kaze No Regret* (1999), an audio adventure game created for the Sega Dreamcast and Sega Saturn. The game included Braille cards, although the release on the Dreamcast also included an optional visual mode for sighted players. Essentially, *Real Sound—Kaze No Regret* was an interactive narrative that resembled a Choose Your Own Adventure style audio book. Other early audio-based games were similarly drawn from text-based narratives that were relatively easy to convert to auditory adventures. But audio-based games have also been developed that rely less on narrative and more on gameplay mechanics. *AudioDoom*, based on the original first-person shooter *Doom* (1993) game, is arranged as a spatial sonic configuration of small environments (Sánchez and Lumbreras 1999). Research into these types of spatially rendered audio-based games has demonstrated that both sighted and visually impaired players are able to conceptualize a physical game space in the absence of visuals. Even without visuals, audio-based games create a mental space in the player's mind that the player can navigate through their mental mapping of the game's environment.

Papa Sangre (2010) is a popular iPod/iPad audio game that relies on binaural audio technology to create a spatial environment in the player's headphones. Without using their eyes, players navigate through five palaces in the land of the dead by tapping on feet on-screen (figure 1.1). The game's story is told only through sound: each castle has its own sonic identity, and different forms of monsters are represented through sound effects. The main goal of the game is survival as the player wanders through the castles collecting musical keys that commonly hide behind one of the many roaming monsters. Monsters respond to player-generated sound, so if players move too quickly or step on an object that makes sound, they

Figure 1.1
Papa Sangre's (2010) interface.

will be chased and killed. The lack of images is one element that makes it more frightening than most games, as players and critics comment:

And even though you can't see anything, or maybe because of it, *Papa Sangre* is terrifying. (Webster 2010)

In fact, every time you hear anything in *Papa Sangre* your heart races, even when it shouldn't. Babies crying, telephones ringing—it's all scary to me now. (Hall 2011)

The pressure & anxiety really teases out the imagination. Real panic sets in when one steps on a bone. Who needs graphics? (re7ox in *Papa Sangre* 2011)

Enjoying playing *Papa Sangre* very much. The most I've ever concentrated while playing a game. (rooreynolds in *Papa Sangre* 2011)

Papa Sangre is great. I played it at the weekend. It reminds me why the radio has the best pictures. (DominicSmith in *Papa Sangre* 2011)

Papa Sangre is a narrative that can be told only through interactivity. It is moving *through* the space that makes it scary. The player controls the pacing, and since walking too loudly can trigger a demon to recognize and chase the player, the player is to some extent in control of the arrival of monsters. Moving too quickly can cause the player to trip, leaving the player open to attack by the monsters. The player must have patience and move slowly through the space. This is not a game that can be rushed. Through controlling the pacing in this manner, the game forces players to listen while they move, paying attention to the sounds that other characters or objects are making as well as to their own sounding bodies in the virtual world. In other words, bodily engagement in the game (both physical and virtual) adds a dimension to the involvement of the player because the game's characters cannot be revealed and the story cannot unfold without that involvement.

The separation of sound from source allows mental imagery to dominate the listener's mind. This mental imagery is a result of our synesthetic experience of sound as a component of a multisensory integration: we typically experience sound in association with image, and thus when the image is not apparent, we might still mentally "see" that image. The sound without image is not disembodied, in other words, because of its corporeal, haptic, and visual associations.

Synchresis: Integrating Sound and Image

The concept of synchresis—that merging of image and sound—highlights the brain's ability to merge distinct sensory inputs. Indeed, our perception of one modality can be significantly affected by the information that we receive in another modality. Most of the existing research into sound and image suggests that, depending on the situation, the modalities could be said to work together (agonistic, or congruent) or against each other (antagonistic, or incongruent) or have no effect at all on each other (neutral). In some form or other, this basic concept has been repeated throughout the literature. Hansjörg Pauli (1976, in Bullerjahn and Güldenring 1994) proposes three categories of music and image—paraphrasing (the music is additive, in that it is congruent with the image), polarization (the music disambiguates the scene), and counterpoint (the music is incongruent and conveys irony or comments on the picture). Musicologist Nicholas Cook (1998, 102) similarly suggests a contradict, contest, and contrast relationship of image and sound. But these interactions can vary depending on context. For instance, psychologist Annabel Cohen (1993) found that music alters meaning in film only when a visual excerpt is ambiguous. Thus, music can disambiguate a situation by establishing a context.

The majority of the available studies on multimodality in the context of audiovisual media address sound's influence on image rather than the inverse. Few researchers have studied the role that image can play in affecting our interpretation of sound. One notable exception is the work of Trevor Cox (2008), who has shown that images can affect the perceived "horribleness of awful sounds." Michel Chion (1994, 69–71) argues that images magnetize sound, meaning that sounds will appear to emanate from an image. In other words, even with two-dimensional images, a ventriloquism effect occurs in audiovisual media (Thomas 1941), and we hear a sound as coming from an image if we associate the sound with that image, even if the sound is arriving from another direction. This ventriloquism effect phenomenon suggests that we bind image and sound together in our minds.

The theory of multisensory integration holds that there is a synthesis or binding of information that occurs between modalities in which the information that emerges could not have been obtained from each modality on its own (Kohlrausch and van de Par 2005). Deborah J. MacInnis and C. Whan Park (1991) refer to this new information as the *emergent meaning*. An example of this phenomenon is the McGurk effect, a perceptual illusion that illustrates how a speech signal can be altered by the visual image of a mouth speaking a different speech signal. Presenting a participant with the audio for /ba/ and the visual of /ga/ produces the emergent meaning of the perception of the syllable /da/ (McGurk and MacDonald 1976). In other words, neither the image nor the sound dominates our perception, but a new emergent meaning is formed.

This idea of a fused audiovisual emergent meaning has been popular in film sound studies. Sound designer Walter Murch remarks, "Despite all appearances, we do not see and hear a film, we hear/see it" (in Chion 1994, xxi). Murch describes a phenomenon that he calls *conceptual resonance*, which occurs between image and sound: the sound makes us see the image differently, and then this new image makes us hear the sound differently, which in turn makes us see something else in the image, and so on. In other words, a new meaning is generated from the ways in which sound and image work together, and therefore analyzing sound separately from image misses out on this emergent meaning. Where the auditory and visual relationship is not direct or causal, the interaction becomes one of "added value" in which "a sound enriches a given image so as to create the definite impression, in the immediate or remembered experience one has of it, that this information or expression 'naturally' comes from what is seen, and is already contained in the image itself" (Chion 1994, 5). Sound can provide an emergent meaning that appears to be inherent in the image but is actually caused by the image's relationship to that sound.

Although it may be possible to separate sound from source in audiovisual media and still obtain significant connotative information about that sound (see, e.g., Tagg 2000), it is worth considering for this discussion contextual signification, whereby the signifiers are dependent on each

Figure 1.2
Tati's *Mon Oncle* (1958).

other for their meaning. The emergent meaning is not contained in the sound or the image individually but is developed through their contextual pairing. For example, the films of Jacques Tati often used highly exaggerated or substituted sounds for comical effect. In the kitchen scene of *Mon Oncle* (1958) (figure 1.2), Monsieur Hulot walks into a kitchen and touches a hot radiator with his hands. The radiator buzzes like a door entry alarm at the touch. We might expect Hulot to articulate some sound at the burning of his hands, but the buzzer fills in both for his exclamation and for the radiator, as if to say "Don't touch" and "Ouch" at the same time. The sound on its own cannot be associated with either the man or the radiator (it does not have any causal connection to either), but through its contextualization, an emergent meaning is formed. Indeed, Tati relies on such sound gags for much of his entire aesthetic: through such contextualization we reexperience otherwise rather mundane scenes.

Although it is common to merge the "wrong" sound with image in today's games (many gunshots, for example, are actually cannons firing and thus completely unrealistic), this emergent-meaning effect was even more apparent in older eight-bit games. Early games often had to rely on synthesized noise for percussive sound effects because sampled sounds were not possible (or sampling was so limited that the sounds became too distorted to sound realistic). The Atari 2600 (also known as the VCS), for example, was a late 1970s and early 1980s console that was very limited in graphics and sound capabilities, so much so that many sprites were highly pixilated, unrecognizable blocks. Outside of the context of the game (often the cartridge or title would explain the game's premise), the images could be seen as largely meaningless. The same meaninglessness occurs with the sound: most of the sounds were harsh square-wave or white-noise blasts that carried no causal relationship to natural sounds and had little resemblance to any known sounds. Nevertheless, put together, the sound and the animated image make sense, and we can hear/see that the short blasts of white noise in *Combat* (1977) are actually gunshots or that the harsh constant rumble is meant to be a moving tank (figure 1.3). Indeed, the sounds of the gunshots and explosions from the tanks are exactly the same as the gunshot and explosion sounds of the helicopters in the same game (although the engine sounds are different).

In addition to sensory modalities altering the perception or meaning of what is witnessed, research has also shown that increasing the number

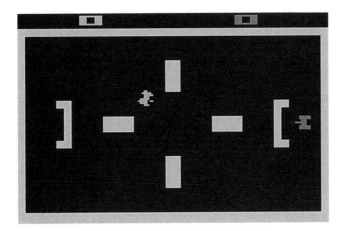

Figure 1.3
Combat (1977): Graphics, animation, and sound imply that tanks are moving around a field with walls, but if sound and graphics were given separately and without the benefit of movement, the viewer or listener would be hard pressed to understand either. Image from *Penny Arcade*, http://forums.penny-arcade.com/discussion/102619/what-were-the-first-games-you-ever-played/p3.

of modalities has different effects on information processing and responsiveness, which can be important to the ability to succeed at a video game. This multimodality research suggests that players may take in information differently when playing a game rather than watching a film: they may retain the information more (or less) and may respond differently to the information. For example, multiple modalities can reduce our cognitive load, particularly if we are receiving complex information. Multimodalities have been successfully employed in helicopter cockpits, where numerous displays and controls mean that pilots must process large amounts of information. Adding haptic and auditory information to the visual controls means that pilots can fly more safely and effectively (Haas 2007). Similarly, in video games, positive-reinforcement sounds (such as collecting swords, for instance) can enable players to determine positive or negative actions in a game more quickly, enabling a quicker learning curve.

Depending on context, multiple modalities can also increase cognitive load. The Stroop task, a popular psychological test of reaction time, illustrates how multimodal interaction can interfere with perceptual processing (Stroop 1935). Here, incongruent information impedes response time, and congruent information improves those response times. The test typically uses a series of words representing colors (such as green, red, and blue). The color of the text is mismatched to the words, so that the word *blue* may be colored red. Although this color-text mismatch is not in a different modality (both are visual), the concept is the same, and the experiment has been replicated with multiple modalities (MacLeod and MacDonald 2000). This concept has also found its way into video games, such as *Brain Age: Train Your Brain in Minutes a Day!* (2005) (also known as *Brain Training*) (figure 1.4) and *Wii Fit Plus* (2008).

In *Brain Age*, the classic Stroop task of identifying the color of text when presented in the opposite color to the word is seen: the word *blue* is written in the color red. The player must say the color "red" to identify the text *blue*. The contradiction confuses the mind, and players often announce "blue" to identify the color. In an interview between Nintendo president Satoru Iwata and game designer Shigeru Miyamoto, Miyamoto discusses his attempts to create an embodied version of the effect (Nintendo 2011):

Miyamoto: We began developing *Wii Fit Plus* with a kind of typical game developer's way of thinking: since the first one has sold well, we should come up with a sequel. It was at that point that you brought up the idea of having multi-levelled exercises which would also train the brain. . . .
Iwata: Ah yes, you're talking about the Stroop Effect.

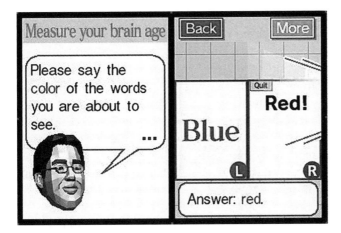

Figure 1.4
Brain Age: Train Your Brain in Minutes a Day! (2005), illustrating the Stroop task.

Miyamoto: The idea was that by working on *Wii Fit* a little more, we might be able to train users' brains as well as their bodies.

Iwata: The Stroop Effect was actually used in *Brain Training* where there was one challenge where the names of the colours didn't match the colour they were written in and the user was instructed to ignore the word and pay attention to the colour. The effect that comes into play at times like that, when you are struggling with two contradictory levels of awareness in order to choose one option, is known as the Stroop Effect.

Miyamoto: Now when you try to do that kind of thing using the body, the example that's easiest to grasp is that of drumming, when the arms and legs are each moving independently.

Iwata: But you did that already in *Wii Music*, didn't you?

Miyamoto: We already did it! . . . That's when I thought that by combining information taken in visually along with holding the controller and the body's balance, we could come up with a lot of fun games. So we began at that point to work on a variety of games separately.

The Stroop effect might be a seldom-used curiosity in game design, but understanding that modalities interact with each other in different ways suggests that multimodal interactions can not only affect the player's perception but can also become an element of gameplay.

Miyamoto hints at the fact that gesture can be treated as another modality that is affected by image (and by extension sound). Indeed, we can make some assumptions about gesture's interaction with sound based on sound's interaction with image. If we accept that an audiovisual binding and emergent meaning is in fact possible between sound and image, we can also hypothesize that a similar emergent meaning could be forged between action and image, action and sound, or all three modalities. In this way, the emergent meaning generated between action and sound would be different from sound alone, suggesting that we do indeed have a different relationship to interactive sound (sound + action, or sound + image + action) than we have to noninteractive sound (sound alone, or sound + image). We may therefore conceive of a kinesonic synchresis,[4] where action and sound can become as equally bound as image and sound and can also lead to an added value in media, where the emergent meaning is thus different from the action or the sound alone.

Kinesonic Synchresis: The Event-Driven Nature of Interactive Sound

Interactive sound in games is kinesonically synchretic: sounds are fused not to image but to action. In other words, interactive sound is event-driven, and the sound is controlled by an action or occurrence that is initiated by the game or by the player. A player-generated event is an event that the player initiates (for instance, by clicking a mouse or by pressing a controller button). Interactive sound in games is primarily the sonification of player-generated events where the player initiates an event and there is a system-controlled (game-generated) sonic response. For example, if I press a controller button, I may hear Mario jumping. The input event is the button press and is controlled by the player. These are direct player-generated events, and the response tends to be immediate. Player-generated events result in what I have previously referred to as interactive sounds. A game-generated event, on the other hand, is an event initiated by the game's algorithms, such as the control of a nonplaying character, a timer-related action, and so on (what I have previously referred to as adaptive audio). Game-generated events may or may not require a response (interaction) from the player. For example, a timer in a game's code may cause an event to happen at a specific point in the game. Player-generated events are always interactive, but game-generated events can be interactive or noninteractive.

Sounds are commonly used as feedback to acknowledge an event. The relationship between the input and output is likely to be synchronous or nearly so[5] and therefore self-evident and predictable. This sets up an expectation on the part of the player that when the same action is repeated, the same (or similar) sound will occur. If I press a button and nothing happens right away, I am likely to be confused. On the Web site *Pretty Ugly Game*

Sound Study, in which users can submit examples of poor game sound, the submissions highlight some of the failures of synchronization and the effect that this has on the player. For instance, discussing *Crazy Frog Racer 2* (2006), one player commented, "The sound effects don't suit the environment at all. It sounds as if it's all happening inside of a room instead of outside. There's also a delay between the sound effect and the movement making the sound" (in Huiberts and van Tol 2007).[6] In this case, there is also an issue with reverberation, but the delay in the event (movement) and sound is noticed by the player and treated as a flaw. If the feedback for an action is out of synch, this is frustrating. The synchronicity of the response helps players to understand the consequences of their action, reducing the learning curve of the game and providing valuable feedback.

The concept of interactive sound as being event-driven suggests that events are repeatable—that if we repeat the action, we will receive the same reaction. This repeatability of events is one of the key elements in sound's ability to provide feedback to the player. Repeatability establishes an expectation that we will hear the same sound as a reaction to the same action. This helps players learn the sound's meaning, increasing efficiency for the players, who can rely on the feedback to help them play the game. Repetition, therefore, can be a useful tool, but repetition also has a downside: players can find it annoying if sounds repeat too often. Most games are designed to be played multiple times, and repeated listening can be tiring, especially if a player spends a long time on one particularly difficult area of the game. Composer Marty O'Donnell (in Vachon 2009, 2) elaborates on this concept in his discussion of the *Halo: Combat Evolved* (2001) score:

The most important feature . . . is that it contains enough permutations and the proper randomization so that players do not feel like they're hearing the same thing repeated over and over. Even the greatest and most satisfying sound, dialog or music will be diminished with too much repetition. It is also important to have the ability to randomize the interval of any repetition. It might be difficult to get the sound of one crow caw to be vastly different from another, but the biggest tip off to the listener that something is artificial is when the crow always caws just after the leaf rustle and before the frog croak every thirty seconds or so. The exception to that rule are specific game play sounds that need to give the player immediate and unequivocal information, such as a low health alarm.

The idea that the same output is always mapped to the same input cannot be assumed, therefore. A sound may be selected from sound files at random and based on priorities or other run-time parameters. A container folder in the middleware engine Wwise, for instance, allows a game's sound designer to group sounds into one folder and tie events to that folder rather than to an individual sound file. A sound file from the folder is selected randomly when the event is triggered. If my character fires a shotgun, I may want the sound of the shells hitting the ground to be randomly selected from a series of six different shell sounds. The first time I press the fire button, I may hear soundfile A; the second time, soundfile B; and so on.

An event may also trigger the run-time generation or selection of a sound based on in-game run-time parameters. A footstep sound may, for instance, be synthesized according to the health of a game character. A wounded character may drag a leg, for instance, or walk more slowly and carefully, requiring different sounds. Vachon (2009, 8) elaborates:

Footstep repetition has often [been] addressed by providing multiple variations of every sample, creating different samples for left and right feet and randomizing volume and pitch. Coupled with the need for multiple surface types, these add up to a large amount of sound effects that need to be managed and carried in memory. The sounds [need] to be similar, because any sample that is too different risks sticking out in a random sequence, so the payoffs of this method are not as big as we could expect. Some audio designers will disassemble their footstep sounds into HEEL and TOE components, making it possible to randomize each component, and to mix and match from different original samples, therefore creating a quasi-unlimited number of possibilities. This yields interesting results, but it is rather cumbersome and is still disconnected with the actions of the character. . . . The ideal system would combine these methods, but link them to the weight and speed of the character. By varying the attack rate of the heel component according to the speed of the character, a simple stationary shuffle would have practically no audible heel component, a walk would have a soft one, and a full on run would have a hard heel. You could even modulate the pitch according to the same parameters. Add a little randomization to pitch, volume and attack rate, and you have an organic footstep system that is never the same but is tied to the behaviour of the character.

Vachon raises several important points here. Sounds need to be connected to the actions of the character. Randomization is only as strong as its ability to tie all of the sounds to the character's action. Randomization increases the believability of the scene, but any sound that is too distinct can call attention to the artificiality of the sound. One of the difficulties with randomization, then, is with the window of variability—the threshold above or below which the sound becomes so different that it draws attention to that difference. A variation that is too different can inadvertently draw our attention to the sound. If a footstep sound suddenly increases in pitch or volume, we may start looking around the game scene for the cause of the change (which may or may not exist), and too much variation can lead to incongruence between sound and image or sound and action. Extending beyond the window of variability—having sounds with too little or too much variation—reduces believability, and the sound becomes incongruent.

Kinesonic Congruence and the Player

Sound in film can be congruent, incongruent, or neutral in relation to the image. In games, however, there is the added modality of the player's events, meaning that sound may be congruent with the image or may be congruent with the action of the player. Sound may equally be congruent, incongruent, or neutral in relation to image or action. A mismatch may occur between the gesture of the player, the imagery of the game, and the sound. Sound designer Kenny Young (2010a) describes how in the game *Heavy Rain* (2010), the sound is tied to the start of the action of the player but not to the image or the gesture of the player:

The speed-dependent dynamic interactions aren't scored effectively with sound. For example, the sliding door in the bedroom does not adapt to the speed that I move it at, there is silence and then a one-shot sound event plays. If I'm moving the door slowly then the result is the sound finishing a second before the [visual] door actually closes. Interestingly, there's a point in the animation where it doesn't matter what speed you are moving at and the door closes at a set speed—this is the point the sound should have played at. Hell, that was probably the idea, but if your system is based on timing rather than designed to deal with the dynamic interaction you are setting yourself up for a fall when the designers come along and change everything (which you know they're going to do). Similarly, when shaving, there is a one-shot "shaving" sound that plays irrespective of how long the razor spends in contact with my skin. I appreciate that getting this right is a bit more work, but given the importance of this mechanic to the game I think this is a bit of a let-down.

Here, the sound is initialized by the player, but there is no congruence between the player's action and the sound: We can refer to this congruity between the player's gesture and the sound as *kinesonic congruity*. With gestural inputs such as the Wii remote, either the player determines the length of the sound output being played (assuming the sound is mapped to the player's gesture), or the sound is mapped only to the start of the event (that is, a one-shot triggered at the start of the input as described by Young above) and the length of the gesture is incongruent with the length of the sound. Sounds are kinesonically incongruent when they fail to map to the action or gesture of the player.

The introduction of the gesture-controlled input devices such as the Wii remote and Xbox Kinect, as well as controllers adapted from real-world objects, such as the musical instruments of *Guitar Hero* (2005) and *Rock Band* (2007), have ushered in a new wave of opportunities for kinesonically congruent sound events. Such gestural control over sounds could provide players with useful feedback regarding the strength of gestures, the length of time that an action took, proximity, direction/angle, and so on. This feedback provides more information to the player, and leads to a better sense of control over the character. Nevertheless, gestural input has considerable consequences for game sound designers. Since the timing of a player's gesture is unknown, the sound designer is suddenly faced with unpredictable sound requirements: how long should a sampled sound effect be if the length of the player's movement is unknown? The response of some sound designers has been to return to sound synthesis. Although it was common to synthesize sounds in the early years of video games, by about 1990 this practice had been largely abandoned, at least for console games, in favor of sampled sounds (see Collins 2008). It was felt that sampled sounds were more realistic in terms of their auditory fidelity and that this realism would drive game sales. Now, however, we are confronted with a need for a different kind of fidelity—kinesonic fidelity. Sound designers have been left to wonder if it is more important to impart a kinesonic fidelity or a sonic fidelity. If I exaggerate my Wii tennis racket swing, emphatically swing it to the greatest of my strength, but hear only a light tennis ball "pop," does it matter that the "pop" was a sampled sound that may be technically realistic if that sound does not map onto my action? And how does this affect the ways in which I play? Do I then adjust my actions to account for the lack of sonic feedback to those gestures? These questions remain unanswered but illustrate the importance of the role of interactivity and embodiment in sonic involvement in games.

I have introduced a number of concepts in this chapter that require further exploration and that suggest that media theories need to be adjusted to account for interactivity. We began by exploring the synesthetic nature of sound in general and determined that sound alone—schizophonic sound—has multimodal implications: it is heard causally and has haptic and visual associations that relate to that causality. When sound is paired with image (synchresis), we might alter that signification in a way that gives sound a new, audiovisual emergent meaning—a contextual signification. This emergent meaning in a sense reembodies the sound, giving it a new causality. We extrapolated this finding to suggest that a similar process may occur when sound is paired with action (kinesonic synchresis), refer-ring to the result as the kinesonic emergent meaning. But what complicates synchresis in interactive media is that the pairing of action, image, and sound varies. In the case of games, it varies to eliminate boredom, to provide more accurate feedback to the player, and to create a more realistic sonic atmosphere. Interactive sounds are not always predictable or obvious in their response to a player's input. The same action does not always result in the same reaction. Kinesonic synchresis is constantly in flux and dependent on a significant number of variables—randomization, parameterization, and prioritization. Given this randomization, the sound-image connection (the audiovisual emergent meaning) and the sound-action connection (the kinesonic emergent meaning) are distinct every time that the game is played, within the window of variability.

The variations in sonic response to players' action are somewhat coun-tered by the window of variability, in which the inconsistency of the sound takes place within a limited range and thus remains within the boundary of both plausibility and expectation. Players expect a certain response but allow for some variations in that response. How does this variability alter multisensory integration? Do players mentally group the sounds together within each window of variability, such that they hear them in the approx-imately same way, with the same meaning, each time that they play? In other words, do players mentally categorize the sonic response into larger abstracted concepts ("shotgun shellness") as opposed to identifying actions as tied to a highly specific sonic reaction ("*that* shotgun shell sound")? If the former is the case, then perhaps emergent meanings in interactive media need to be examined by studying gestalts rather than specific indi-vidual examples.

We might also suggest that this gestalt meaning influences the ways in which players hear sound. Thus, a highly synthesized sound effect paired with the swinging of a Wii sword may lead to an emergent meaning of sword fighting, even if the sound on its own is not realistic (perhaps a blast of synthesized white noise rather than a sword sample). As with the examples of audiovisual emergent meaning discussed above, the combina-tion of action and image may lead to a new kind of fidelity of the gestalt of the action-image-sound that means that players can forgive the failings of realism in one of those modalities. It may feel more real to the player to have kinesonic congruity than high auditory fidelity.

Our interaction with sound thus adds significant implications for our understanding of sound in media. Through multiple modalities, interac-tive sound may encourage a three-way emergent meaning where new meanings are created through those interactions. Therefore, an emergent meaning among sound, image, and haptics may be different than between sound and image alone. If this interactivity makes players hear sound in a different way—that is, in its kinesthetic, haptic, and visual context—how does this change players' relationship to the media? Are they more or less involved in narrative and characters? Do they create a stronger bond with those characters, given that they are directly involved in their sonic activity? I explore these questions in the following chapter.

2 Being in the Game: A Sonic Approach

In chapter 1, I showed how interactivity adds a new dimension to meaning in media: sound is contextualized in terms of its haptic and visual associations, each of which may bind to sound to create a new emergent meaning. But these contextualized meanings are not exclusive to interactive media: sound carries connotations of its haptic and visual associations even in the absence of these modalities in media. What is unique about interactive sound is its ability to create new associations through its haptic recontextualizations—a kinesonic synchresis. The embodied cognition theory of the mirror-neuron system in the brain provides evidence that illustrates the ways in which a three-way mapping occurs among sound, image, and action. Research has shown that the same group of neurons in the brain fires when either performing or observing an action, and so these are referred to as *mirror neurons*. Our brains mirror the action of what we witness as if we are performing that action ourselves. In other words, our emotional and neurophysiological states can be directly affected by what we see or hear: if we see pain or fear in someone else, we understand this in terms of our own physiological experience of similar pain or fear (Niedenthal 2007). For example, researchers at the University of California at Los Angeles found that the neurons that normally fire when a patient is pricked with a needle will also fire when the patient watches another patient being pricked (Ramachandran 2009). Mirror neurons are therefore closely tied to our experience of empathy. Neuroscientist V. S. Ramachandran (2009) believes that mirror neurons dissolve the barrier between self and others and so refers to them, with humor, as "Gandhi neurons."

Research into mirror neurons has found that the same neurons fire whether an action is performed, seen, or heard (Kohler, Keysers, Alessandra, Umilta, Gallese, and Rinolatti 2002). In other words, when we hear a sound, our brain responds as if we are also seeing and experiencing the action that is creating the sound. We understand the actions of others because we mentally mimic those actions visually, sonically, and gesturally (Keysers et al. 2004). Moreover, the neuronal response is much stronger if we have undertaken the action before: we re-create our previous experience of those actions mentally. In other words, action-related sounds are associated with an image and an action in our minds: as discussed in the previous chapter, we are always listening causally, and sound is always an embodied, multimodal experience.

India Morrison and Tom Ziemke (2005) argue that this mirror neuronal response is not limited to our witnessing other human beings but also is experienced when we observe virtual characters and thus "can facilitate a user's identification with the character's 'body' as well as provide the groundwork for empathy" in games by drawing on mirror neuronal theory. A cognitive multimodal mapping occurs between visual, motor, and auditory representations that is closely integrated to our own feelings of empathy, even with virtual characters. Through multimodal integration in our brain, we might identify with game characters.

The word *identification* in this context refers to feelings of affinity, empathy, similarity, and liking for a character by an audience. A variety of theories have arisen as to our identification with on-screen characters. Cassandra Amesley (1989), in discussing television, suggests that a "double viewing" occurs when we watch a program in which the characters are simultaneously both real and constructed in our minds: we project and transfer our own beliefs about the character onto that character. Thus, the character becomes us as much as we become them, in a blurring of real and imagined personality traits. Through empathy and mental role-play, we adopt the character and can extend our sense of self into the character. With video games, James Paul Gee (2004, 55–56) proposes that rather than a dual character, three simultaneous identities occur during gameplay—the player (the real world), the character (the virtual world), and the projective identity, which is "the interface between—the interactions between—the real-world person and the virtual character." This projective identity, argues Gee, is a combination of the character and the player's belief (projection) about the character's personality. Each of these theories relies on the assumption that players create an extended sense of self through their identification with on-screen characters. With television, this extension is entirely virtual, but video games offer a physical extension of the self through the use of controllers.

Extension and Incorporation

Game controllers can become an extension of the body into the virtual world. Scientific research has shown that the sensorimotor parts of the brain will respond to tools like game controllers as a part of the body—an extension of a hand or an arm (Morrison and Ziemke 2005). This extension is strengthened by players' ability to adapt to the controller as an input device, and as we increase our expertise using the controller (thus reducing the feeling of mediation), the gap between person and avatar is reduced. The more adept that players become at using the device, the less they notice the controller, and the more engaged with the game they can become (Cleland 2008, 222). For example, when learning to drive a car, we pay attention to the many tasks at hand—the pressure of a foot on the pedals, the placement of hands on the steering wheel, the shifting of gears—but as we become more experienced, we do not notice these tasks, and they become second nature to us. This extensibility of tools is not a new conception. In 1945, Merleau-Ponty (1998) wrote at length about how tools can become an extension of the self. A frequently used tool becomes a part of us because we no longer focus on it as a part of our experience: it facilitates a feeling of nonmediation between us and the (in this case virtual) world. Marshall McLuhan (McLuhan and Lapham 1994, originally 1964) spoke extensively of the concept of technologically mediated extensions of mind and body, arguing that any media form or technology can become an extension of our senses and therefore ourselves. Philosopher Don Ihde (1979, 508) also expresses a similar concept in his discussion of embodiment relations, where "the experience of one's body image is not fixed but malleably extendable and/or reducible in terms of the material or technological mediations that may be embodied." Our sense of self, in other words, is mediated by our technologies.

The rubber-hand illusion illustrates how multisensory perception can influence this sense of self and bodily boundary. In this series of experiments, researchers found that by stroking a rubber hand in view of a participant while simultaneously stroking the participant's real hand (hidden from view), the participant takes "ownership" over the rubber hand to the extent that they can "feel" through the rubber hand: "The effect reveals a three-way interaction between vision, touch and proprioception, and may supply evidence concerning the basis of bodily self-identification" (Botvinick and Cohen 1998, 756). Although experiments that involve the rubber-hand illusion have not used auditory feedback as an experimental condition, such work illustrates that the senses are integrated into the brain and that multimodal feedback affects the ways in which we understand our sense of self and other. If haptic (touch) and visual feedback can lead to feelings of bodily extension, using image, sound, and haptic feedback (including proprioception and gesture) as sensory input in a virtual world (as in a video game) could lead to a similarly altered sense of self. If we obtain sonic and visual feedback from a virtual space and are physically involved in that space through our own gestural input, then after we overcome the feeling of mediation of the controller, we are likely to extend our sense of self to our on-screen persona.

The rubber-hand illusion demonstrates that we have a body schema into which we may assign noncorporeal and corporeal objects. To the extent that we incorporate such objects into our body schema, however, these must be representational of our own body: if we replace the rubber hand with a wooden stick, the effect fails (de Preester and Tsakiris 2009). Consequently, there is a distinction between incorporation and extension. The rubber hand is incorporated into our body image and becomes a physical part of our body to the extent that the brain takes ownership over it and our body schema is altered. Extension, on the other hand, is the feeling of nonmediation between the self and the world through the use of a tool or technology. Incorporation and extension are closely related. Helena de Preester and Manos Tsakiris point out Merleau-Ponty's conflation of the terms in his discussion of the blind man's cane, in which the cane becomes a tool through which the blind man senses the world. There is an important distinction, however, in the sense that with incorporation the object is brought into our existing body schema, and with extension the object becomes an extension of that body schema into the peripersonal space.

The peripersonal space is an intermediary space between our body (personal space) and our view of the external environment (extrapersonal space) (Cleland 2008, 252). Lucilla Cardinali, Claudio Brozzoli, and Alessandro Farnè (2009) argue that auditory information exists in this peripersonal space and that sound therefore is an extension of our sense of self rather than an incorporation into our body schema. In this way, sound in games could be said to extend our sense of self beyond our physical body and into the intermediary space between ourselves and the virtual world or into the virtual world itself. Sounds that we make—including in the virtual world—become a sensory extension of our self into that virtual world. The auditory realm of games thus becomes an extension of the self, a technological body through which we experience the game world. Sound, in other words, has a unique ability to extend the self into virtual space.

The elasticity and mutability of our body schema are readily apparent. As Don Ihde (2002, 138) comments: "We are our bodies—but in that very basic notion one also discovers that our bodies have an amazing plasticity and polymorphism that is often brought out precisely in our relations with technologies. We are bodies in technologies."[1] The ease with which we adjust our body schema through sensory perceptions illustrates how players may be able to identify with a game character as an extension of the self through sound. In other words, it is partly through sound that players become a character. In this way, the body is extended not through the controller but through sonic interactivity with the game character. The character is the tool through which players experience the virtual world. An important consideration of this sonic extension into the game is the idea of self-produced sound.

Self-Produced Sound

Self-produced sound can be defined as sound produced by one's own body or bodily movement (Ballas 2007). Sounds that are self-produced provide us with important feedback about the world in which we live and help to delineate our sense of self. From the earliest moments after birth, we experience perceptual events that identify the self versus the other. Some scientists have studied self-produced sound in infants at length. In one series of studies, Philippe Rochat (1995) used a variety of methods to test auditory feedback on infants and found that infants sucked their pacifier differently based on the auditory feedback that they received. By age two months, infants could identify kinesonically congruent or incongruent auditory feedback based on their own actions. Rochat (1995, 397) found that "when infants cry, the sound they hear is combined with kinesthetic and proprioceptive feedback. This intermodal combination is uniquely specifying the perceived self. Sounds originating from another person or any other objects in the environment will never share the same intermodal invariants. By the second month, when infants start to vocalize and to babble, they appear to explore systematically the specificities of their own voice and the potentials (or affordances) of their own vocal track." From early in our lives, we use sonic feedback to determine whether something is a part of ourselves or external to ourselves. Our connection between the physical action and the sonic reaction is much stronger when it comes to self-produced sounds (as opposed to externally produced sounds), due to the embodied connection between self and sound. In other words, we have a physical experience of sounds that we produce ourselves that is different from other sounds in our environment, and we use these sounds to help to delineate our body schema.

This connection to self-produced sound suggests that when players produce the sounds in a game (in the sense that they are immediately receiving feedback for their own actions), they are experiencing those sounds cognitively as "their" sounds. Because players receive immediate feedback tied to their own proprioceptive or kinesthetic actions, the sounds become a part of self rather than other. In this way, sound helps players to become a character, or perhaps more accurately, their character can become a part of their sense of self. Kinesonically incongruent sounds are therefore more likely to be particularly disturbing.

In summary, the delineation of our own body in space is a somewhat subjective phenomenon that is subject to change through reconceptualizations of our body schema and the peripersonal space. The peripersonal space serves as an auditory intermediary space between the player and the game world. Self-produced sound in the players's peripersonal space is one important agent through which they extend the body into the virtual world. Put simply, if players undertake an action and have a kinesonically congruent sonic response because that sound takes place in their peripersonal space and because they caused that sound, then that sound becomes an extension of their body schema into the virtual world. Sounds that they make and sounds that their character makes exist in the same sonic peripersonal space and are often kinesonically congruent. In this way, a player can become the character, or a character can become the player because they exist through extension in the in-between space sonically.

Sonic Game Space: Point of Audition in Games

In the field of game studies, the concept of a separate space in which gameplay exists has drawn on the theory of the "magic circle" first proposed by Johan Huizinga (1955). The magic circle is "shorthand for the idea of a special place in time and space created by a game. . . . As a closed circle, the space it circumscribes is enclosed and separate from the real world" (Juul 2006, 164). The magic circle is a kind of psychological space in which the game exists, and it takes place in the space immediately around us as we play. It becomes a separate zone in which the virtual world dominates. But arguments have developed within game studies as to the nature of—and the existence of—the magic circle. Some have noted that the magic circle is not definitive: public performance of game playing and the notion of spectatorship suggest that an exclusive magic circle cannot exist. Moreover, in massively multiplayer games such as *World of Warcraft* (2004), the concept of a magic circle becomes problematic because demarcations of space become more soft and fluid when players role-play their character offline at fairs or conventions that are outside the context of the game (Lammes 2008).

The literal view of the magic circle as a physical space—one that is "quite well defined since a video game only takes place on the screen and using the input devices (mouse, keyboard, controllers), rather than in the rest of the world" (Juul 2008, 49)—is overtly ocularcentric. Michael Nitsche's (2008, 3) description, for instance, relies on narrowly conceived notions of games as strictly visual phenomena: "The screen remains an important layer as it is mainly through the screen that the game worlds can unfold and become accessible to today's player." As with other aspects of games, a focus on space in game studies has to a great degree relied on the visual space of the screen to the exclusion of our other sensory interactions with that space (for exceptions, see Stockburger 2003 and Grimshaw 2007).

To some degree, this conception of game space is borrowed from cinematic studies of the frame of the cinema or television screen. Both Lev Manovich (2001) and Anne Friedberg (1993) conceive of the frame of the cinema screen as a window to "the existence of another virtual space," a "space of representation" in which "the viewer simultaneously experiences two absolutely different spaces that somehow coexist" (Manovich 2001, 95). The visual on-screen objects cannot be directly or meaningfully interacted with, and thus "The metaphors of the frame and the window both suggest a fundamental barrier between the viewer and the representational objects seen in the image-screen" (Cleland 2008, 171). Manovich (2001, 108) explains that "classical cinema positions the spectator in terms of the best viewpoint of each shot, inside the virtual space. This situation is usually conceptualized in terms of the spectator's identification with the camera eye. The body of the spectator remains in her seat while her eye is coupled with a mobile camera." But such suggestions fail to recognize the role that sound (and embodiment) can play in our phenomenological experience of the construction of space. Audiovisual media do not only take place on the screen: such media simultaneously take place in the auditory, peripersonal space around us.

The three-dimensionality of visual game space means that to some extent the visual frame of the screen is no longer the physical boundary of the space. We know that in most games we can move into the off-screen space, and thus we have a mental conception of the space as being much larger than what is constrained to the screen. In this way, "Through intangibility beyond the depicted space, the virtual camera becomes a simulation of 'I' rather than 'eye'; a simulation of viewer-derived presence in space rather than an anthropomorphically based viewing apparatus. The animated virtual-camera simulation of 'I' is simultaneously both a party of the scenic composition and beyond it" (Jones 2007, 228). As Mike Jones (2005) describes, "There is an unvoiced acceptance on the part of viewers that all that is important in a scene will take place within the screen's frame. But in the 21st century, many of the key aesthetics of audience acceptance and visual understanding of a broader cinematic space derive not from cinema but from computer gaming. A larger, more complex imaginary world composed by an auteur in-space rather than in-frame." This abandonment of the frame in favor of a player-controlled camera in the virtual three-dimensional space of many modern games is part of a larger macro-*mise-en-scène* or *mise-en-space* (Jones 2007, 227).

The *mise-en-space* of games means that the borders of the screen space are not the borders of the viewable space. The player in many narrative 3D games has control over the camera and can turn around to view what was previously off-screen, reducing (if not eliminating) what is often called in cinema the "exit-sign effect." Named by film sound designer Ben Burtt, the exit-sign effect refers to a phenomenon that relates to discrete sounds that are placed in the rear surround loudspeakers. Pans of fast-moving sounds to the left or right of the screen boundaries lead the viewer's eye to follow the sound through the acousmatic (off-screen)[2] space and toward the theater's exit signs at the side of the cinemas. Some listeners argue that in games in which players can turn their characters around to see what caused those sounds, such discrete sounds in the rear surround speakers can take the player out of the immersive experience. As one player describes, "Without fail, during a gaming or movie session the first time I hear a sound come from a rear speaker, I'm actually brought out of the gaming or movie experience. If you turn around, you don't see the person the sound came from. You see the speaker or a wall or the plant your grandma gave you. So you have to remember that the next time you hear a rear channel sound. After a few minutes of playing the game that becomes less of a problem, and it can really be helpful to know when a bad guy is shooting at you from behind. But the fakeness is still in the back of your mind, and you just know it doesn't sound quite natural" (Sayre 2008). As with adapting to the awkwardness of a game controller, we may similarly adjust to other limitations of a virtual space.

Typically, the field of on-screen view in a first-person-perspective game is between 65 to 85 degrees (Stevens and Raybould 2011, 90), meaning that the majority of the virtual world (that is, the other 275 to 295 degrees) takes place through the sound off-screen. The construction of the virtual world as a three-dimensional space therefore relies for the most part on sound—much more so than in cinema, which tends to have a much wider field of view. Mark Grimshaw (2007) and Axel Stockburger (2003) both refer to the potential of players to deacousmatize sounds by adjusting their position in the game (either through the virtual camera or through player movement). The player can alter the screen view by turning toward a sound and following it into that off-screen space. Grimshaw (2007, 192) points to the ability of game designers to make this deacousmatizing ability an important component of gameplay, particularly in first-person perspective games. *Dead Space* (2008), for example, is a survival horror first-person shooter game, that—perhaps to some extent because it lacks an on-screen heads-up display that is common to first-person games—makes significant use of acousmatic sounds. In several places in the game, sound is used to bait the player to investigate an off-screen area. The creative director Wright Bagwell describes it (in Graft 2011): "Okay, when you come into this room, you're going to hear this banging sound, and there's going to be a light flashing in the corner, and we're going to try to lure the player over because his eyes are going to be drawn to the flashing light or the sound of the thing or the spinning thing in the room. And then right when he's walking over to it and he's looking at this thing that we're baiting him with, we'll hit him from the side, or we'll reveal it to be something else." The use of sound was so successful that the sound director Andrew Boyd (in Turtle Beach 2011) continued to use sound in this manner in the sequel, *Dead Space 2* (2011):

Surround [sound] really brings the world to life. We made the game so the stereo mix would also sound great, but there are a couple of key advantages to experiencing it in 5.1. The first is immersion. The ambience and atmosphere of the game is so much more effective when it can fully envelop you. It's one of those specific things that we can do in audio that the graphics guys can't—they're stuck on the screen, while we're right out in the room with you, which can really help with the tension and horror in a game like *Dead Space 2*. A subtle, creepy sound will come up in the surrounds, and I've seen players not just turn Isaac around but turn around in their own chairs to see what might be there. Obviously this kind of effect is available in any surround presentation, but in interactive horror it's particularly effective.

As Boyd describes, sound has an advantage over visuals because of its unique ability to extend beyond the screen into the player's space. Players may turn around in their chair because the sound occurs in *their* space, not on the screen in front in the character's space. Because players can deacousmatize the sound and that sound can exist in a three-dimensional peripersonal space, the sound extends the game space well beyond the frame of the screen. In other words, space in games is not created by the visual frame in the same way as it is created in cinema (without the audience's ability to change viewing angle or explore off-screen space). Instead, space is often dominated by spatialized sound in a way that cinema is not, and only through the interactivity of the player is the space of the virtual world fully realized. Interactive sound serves to encourage that exploration of the world and reinforce the sense that there is a much wider space to uncover.

The peripersonal space of video game play in which the sound takes place is even further extended and reinforced by Wii controllers. The Wii remote has a small speaker inside, which allows the extension of auditory space of the game even closer to the players' bodily actions. In other words, although sound may surround the players, the Wii remote—a bodily extension—allows self-produced sound to generate from the players' body in the game. For example, in *The Legend of Zelda: Twilight Princess* (2006), players use the remote as a bow and arrow. As they stretch back the string of the bow, the sound emanates from the remote, which is close to the body. When players let the string go, the sound travels back to the television/surround speakers as if the virtual arrow has left the body's touch and traveled into the game space. This perceptual depth can become even more important with colocated multiplayer games: rather than have sounds merely emanate from the television/surround speakers, the players have their own dedicated sounds that emanate from the remotes. In this way, the players are more clearly delineated but also have their own sonic signature in the game.[3] In all games, but especially with the Wii remote and contemporary embodied gestural inputs, sound becomes that extension of the game into the world—and the world into the game.

It is worth exploring the audiovisual representation of space in games in greater detail to highlight the particular role that sound plays in constructing the space as well as positioning the player in that space. In cinema, the auditory perspective is often created by a careful positioning of the microphone to blend direct and reflected sounds[4] so that they duplicate the real space in which the acting takes place in a way that the sound perspective matches the visual perspective. The microphone mix typically mimics the angle and distance of the camera (Belton 1985, 68). Although there are many cases where the auditory and visual perspective do not match (such as with long shots paired with close-up sound), usually the camera and sound reinforce each other and create a sense of distance between the audience and characters on-screen. Perspective in games, however, is fundamentally different from perspective in films because the player usually has some control over both the visual camera and the auditory perspective. In a game, a player can explore a strange fountain sound by moving closer to the source, which gives players some auditory control over their space. In games, both the actual position of the emitter in the virtual space and the distance from the listener must be taken into consideration to create the auditory perspective, which occurs through attenuation (natural weakening of the signal tied to distance) and spatialization (position within the three-dimensional world). This perspective is important in shaping the player's perception of the game and provides the means by which the player engages with the game character or objects.

In today's video games, the first-person and third-person points of view (including third-person trailing view and third-person omniscient view) are most common, although point of view is by no means fixed. Many games today switch perspectives in cinematic sequences to give the player control through camera movement or to meet the requirements of a scene where a particular perspective may be too awkward. Aki Järvinen (2002, 116) refers to *point of perception* rather than *point of view* in games, meaning the visual and auditory position from which the player perceives the game. However, this assumes that the perspective is the same between the auditory and visual realms. In many cases, there are multiple simultaneous points of perception, and they are not always synchronous between point of view and point of audition. For example, the sound does not necessarily map from the character's viewpoint to the player's viewpoint. Players hear the game world from their own first-person perspective: objects on the right-hand side of the screen make sounds in the right-hand side speakers, whereas the right-hand side may be the character's front. For instance, in most side-scrolling platform games, the character moves from the left-hand side of the screen and faces the right. Sounds occurring in front of the character occur to the right side of the player (figure 2.1) (see Rambusch 2005 for a similar point about movement and controllers).

The first-person point of view positions viewers as if they were inside the frame and looking through a virtual camera, which allows a close match between the point of view and point of audition (see below)(figure 2.2). Many authors have alleged that the first-person point of view is the most immersive, since this perspective is close to a natural subjective viewpoint. Mark Grimshaw (2007, 20) suggests that first-person perspective's advantage over other perspectives is that it can use the camera to provide

Figure 2.1
New Super Mario Bros Wii (2009). If players were hearing from Mario's perspective, the sounds of the ship should occur in front of them rather than to the right. Image from *MundoGamers*, http://www.mundogamers.com/wii/imagen/3050/new-super -mario-bros-wii.html.

Figure 2.2
First-person perspective. There is no visible player-character, and players move through the space as if they were holding the camera.

information about the character's state, as in *Deus Ex* (2000), where if the player character gets drunk or poisoned, the camera will "shake and become wobbly." Laurie H. Taylor (2002, 20), however, argues that first-person perspective does not feel more natural or lead to greater immersion, suggesting instead that through its limited viewpoint it "impoverishes spatial presentation in the game and removes the possibility of the player playing within the game space, which removes the possibility for the player to internally experience the game space. The presumptive 'consistency' of visual representation of the first-person point-of-view neglects the heterogeneity and complexity of visual representation and perception in the actual world." Despite the reduced field of view that often is found in first-person perspectives (as noted above, 65 to 85 degrees), spatial audio cues can compensate for some of this visual loss by providing audio beacons that draw players to the off-screen (Grimshaw 2007, 148) or simply signal the presence of off-screen space.

In third-person point-of-view games, players see their character on-screen, although other areas of the screen may (or may not) remain in first-person perspective (figure 2.3). In some cases, as in *Legend of Zelda: Ocarina of Time* (1998), players see the world "over the shoulder" of the character, which is known as third-person trailing perspective. Some games,

Figure 2.3
Third-person trailing perspective. There is a visible player-character on-screen, and players view the world from behind or beside that character. Image adapted from Google 3D warehouse using Google's SketchUp software taken from the "Grand Theft Auto San Andreas Grove Street" model by "Hippy."

like *Kinect Adventures* (2010), have a semitransparent third-person trailing perspective, so that players can see through the character. In other games, players can see only over the shoulder of the character, and in still others, players can turn the character around to face various directions. In this way, players can view much of the game world in the immediate vicinity of the player-character at any one time, including directly behind the character, in a "contextualized presence, so that the player can experience the space through the player-character as other than simply a geometric construction" (Taylor 2002, 28). Third-person perspective can lead to greater extension into the game space because it allows players to navigate the space easily because with a wide field of view and an on-screen character that helps to gauge size, players can see how high a wall is, what is around the corner, and so on.

In third-person omniscient visual perspective (also known as "God view"), the player typically has a top-down or isometric view of multiple game characters. Because they are not always controlling a single character in the game, players may switch between characters, simultaneously play multiple characters, or control the overall environment (figure 2.4). The player is usually able to control the camera to zoom in or out or inhabit a character for a time. The downside of third-person omniscient point of

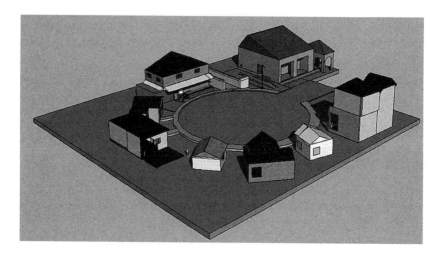

Figure 2.4
Third-person omniscient perspective. There is a visible player-character on-screen, and players view the world from above that character. Image adapted from Google 3D warehouse using Google's SketchUp software taken from the "Grand Theft Auto San Andreas Grove Street" model by "Hippy."

view is that players do not directly identify with any one specific character: they are external to the game world with no direct space inside the narrative (Taylor 2002). However, such perspective has advantages, too. Artist Brian Cullen (2010, 91) suggests that omniscient perspectives are popular in all media and that "In cinema and television, we are invisible. We observe people who cannot observe us. We hear all; no relevant conversation gets past our ears. . . . Technology facilitates a power to transcend scale and physical boundaries. As the camera moves from home to home, penetrating doors, walls, ceilings and floors, the audience assumes the power of immateriality in an imaginative manner." Players have a similar omniscience in third-person games, experiencing both intimacy and distance with the game's characters.

Some theorists, such as Steven Poole, have argued that third-person perspective is less immersive than first-person perspective since players' point of view is disembodied by their ability to see a character on-screen (Poole in Taylor 2002, 133). But as Taylor (2002, 10) argues and as is described in detail below, Poole here relies on vision as the only means of embodiment: "Poole thus equates mere optical accuracy or verisimilitude with the visual presentation of embodiment in the game space." Likewise, some argue that seeing the character that we are playing on-screen is less immersive and more unnatural than first-person point of view, but it can be argued that "a different-looking body would not make me play differently. When I play, I don't even see her body, but see through it and past it" (Aarseth 2004, 48). The character, as an extension of the self, becomes as unnoticeable as the controller as players become involved in the game. Whether or not the player character is semitransparent, players may eventually not pay attention to the on-screen character, since the character becomes a tool through which they act, just as they would a controller, and thus the mediated sensation disappears as the character's movements become familiar.

We can draw on mirror-neuron theory to suggest that third-person perspective may function similarly to first-person perspective in terms of the player's identification and empathy with the character, since players convert information about others into an egocentric frame of reference. In other words, the ways in which we experience a character are always in some sense in first-person. Viewing the player-character in third-person perspective is similar to observing the behavior of individuals in everyday life: we interpret actions in relation to our own first-person terms. We may transform what we see into body-centered information relating to spatial awareness. The location and sensation of our body parts and basic

emotions and our mirror neuronal mappings provide empathy and identification in both first- and third-person perspectives.

Despite the many different points of view (and here I cover only the
currently most common), the auditory position of the listener-player is
often the same, and sound in games is used "to enhance the feeling of
immersion rather than to reinforce the viewpoint of the camera" (Rowland
2005, 7). The auditory perspective has been referred to in cinema as the
"point of audition," although it may be more accurate to regard this as a
zone of audition (Chion 1994, 89–92). Our auditory perspective is created
through a variety of means, including sonic envelopment, proxemics, and
spatial sound.

One of the key functions and effects of sound in games is to immerse
us in the virtual world through the sense of sonic envelopment. The term
envelopment in relation to spatial sound has been used to describe sometimes overlapping and even contradictory concepts. Jan Berg (2009) sum-
marizes some of the ways in which the term has been used in the scientific
literature to refer to the sense of sonic spaciousness, the subjective immersion of the listener, the fullness of sound images around a listener, the sense
of being enveloped by reverberant sound, and the sense of being surrounded by sound. Here I define *envelopment* as the sensation of being
surrounded by sound or the feeling of being inside a physical space (enveloped by that sound). Most commonly, this feeling is accomplished through
the use of the subwoofer and bass frequencies, which create a physical, tangible presence for sound in a space. Whereas image lies external to us,
sound is an extension of ourselves and also physically permeates us. The
subwoofer shakes our material body when it rumbles with the sound of an
explosion, and the breeze from the loudspeaker is palpable and tangible.
Sound waves are both acoustical and tactile, involving our bone structure
and body cavities: we hear through both our ears and our bones (through
bone conductance), and the sound resonates in our body. It comes as no
surprise, then, that envelopment has been shown to be important to listeners in creating a sense of presence (Rumsey 2002)—that is, in reducing
the sense of mediation between audience and virtual space. Sound thus
is responsible to a significant extent for the sensation of being present in
a virtual environment (Berg 2009).

Proxemics is the study of the distance between people as they interact—
the personal space that we all maintain. Auditory proxemics thus relates
to the distance between recorded presences and the listener (Moore,
Schmidt, and Dockwray 2011). Close miking, for instance, can give the
illusion that we are closer to the speaker than if the microphone is at

a distance from the speaker's mouth when recorded. In games, proxemics can create a distance between the player and the characters speaking to that player, which helps to create a sense of space. In games, vocal proxemics can create the illusion of someone shouting from a distance or whispering in our ear. The spatial positioning of sound effects around us also helps to represent the sonic environment of the visualized space. This spatialization is accomplished using (assumed) loudspeaker or stereo headphone positioning (mixing),[5] digital signal-processing effects that simulate a space (such as reverberation), and rendering based on the position of the player-character and sound emitter (such as occlusion and diffraction effects, where objects in the space obstruct and alter the sound between the emitter and character/player).

The auditory perspective is thus constructed so that the player's perspective is that of the main character, with objects in the virtual space sonically represented and "transplanted" into the real space. The three-dimensional visual space of the game becomes a real three-dimensional sonic space, with approximated sizes and distances replicated, in some cases, into real sizes and distances in the player's space.

In games, the auditory perspective is always the same in the sense that the sonic subjectivity—the player's ability to make sound and to experience the sound as if in that virtual space—remains regardless of point of view. Players hear sounds in the character's space in a way that is analogous to first-person point of view, but they also hear sounds that are external to the game's diegesis and that are unheard by the character (interface sounds that are analogous to third-person point of view). Even when players move the camera, they may remain at least partially in the same auditory space. In most racing games, for example, there are multiple camera angles. The player can select the helmet cam view, a view from the back of the car, a more zoomed-out version of that angle, and a camera angle from the front bumper of the car (on the hood) (figure 2.5). Despite the many available viewing angles, the audio in many racing games remains the same regardless of visual perspective.[6] Players hear sound as if they are in the car even when they are visually above the car, behind the car, and so on. Logically, though, behind the car they should hear a lot of exhaust sound (particularly since these are race cars) and no interior sounds. Even in cases where games adjust the engine sounds based on the viewpoint, players are often still given the interior sounds as well. The auditory position remains as if the player is in the driver's seat driving the car regardless of visual position. In *Need for Speed: Shift 2 Unleashed* (2011), for example, surround sound is used to create a sense of immersion

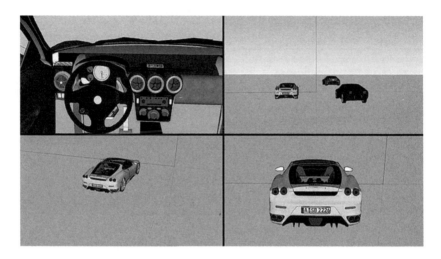

Figure 2.5
Four common viewpoints of a race car. Top left: first-person inside with controls
(helmet cam); top right: first-person through the windscreen; bottom left: third-
person overhead right or left; bottom right: third-person directly behind the car.
Images adapted from Google 3D warehouse using Google SketchUp model with cars
from "Anonymous" and "Jimo."

through hearing both internal and external car noises. Players can hear
the car tires rolling, the gears shifting, and the brakes squealing even when
they are visually positioned outside the car. The sound situates players in
the driver's seat, and the constant noise envelops them as if they were
inside the vehicle. Racing games are the most obvious example of this
auditory positioning, but the effect exists in most games. Players are primarily staged in one auditory position as if they are the main character,
and they sonically experience the virtual world as if they are in the middle
of it, not in front of it on a screen.

Spatial Sonic Embodiment in the Game

A few research studies have shown sound to have an important effect on
the degrees of immersion that are experienced by the player in games. Scott
D. Lipscomb and Sean M. Zehnder (2004) tested the absence and presence
of the composed musical score in the game *The Lord of the Rings: The Two
Towers* (2002). Using a quantitative analysis, the results showed that music
contributed to the sense of immersion in the game. Lennart E. Nacke, Mark
N. Grimshaw, and Craig A. Lindley (2010) studied sound and immersion

in the first-person shooter genre by using sound-on, sound-off subjective and objective tests, arguing that sound effects contribute to immersion. Kristine Jørgensen (2008) and Alexander Wharton and I (2011) have also shown that sound influences both the functional aspects of gameplay and the emotional connection to the game world. These prior studies indicate the importance of sound to the feelings of immersion and connectedness to a video game. But although they illustrate that sound influences immersion, these studies have not examined why or how sound influences immersion.

With kinesonic congruence, players have a sonic reaction that matches the action that the body is making, thus embedding personal expressiveness into the game through the character. Self-produced sounds are one means by which players extend the self into the game space. Even in cases where players do not have kinesonic congruence with sound, however, they still receive some of that sonic response to their actions. Although the physical body did not kinesonically create a sound, mirror neuronal research and the mimetic hypothesis (discussed in the next chapter) suggest that players may still feel that it did through mental or corporeal imitation. If we hear sound in terms of our own embodied experience of that sound, then when players hear those action sounds in games, even though they did not kinesonically create them, they may hear them as if they created them. In other words, players have a direct, embodied interaction with the sounds that they evoke and hear in games, and coupled with physical or kinesonic-congruent action, these sounds (and thus the game character) can become an extension of the self.

The pairing of action and sound suggests that players may also come to hear interactive sounds in a new way. If we understand sounds in terms of our previous experience of creating similar sounds, how is this understanding disrupted when players create sound without the same embodied actions that normally are required for such sounds? For instance, if prior to playing a game, I have experienced the sound of a baseball hitting a bat when I have played baseball, then how does my embodied understanding of that sound change when I press a button (or swing a virtual bat) to make the sound of a baseball hitting a bat? Will this new method of evoking sounds—making them virtually rather than creating them directly—alter the ways in which I experience and understand sound?

In this chapter, I have introduced some ways in which sound increases the immersion of the player and aids the identification with the main player-character. If the player is always in the first-person auditory perspective, as my argument above posits, sound has a considerable effect on immersion and identification. Sound exists in the peripersonal space, extending an intermediary space between ourselves and the virtual world. Rick Altman (1992, 60–61) writes similarly of film when he states that sound "constitutes the perfect interpellation, for it inserts us into the narrative at the very intersection of two spaces which the image alone is incapable of linking, thus giving us the sensation of controlling the relationship between those spaces." Sound thus helps to bring the game space into the lived space.[7] Sound reconciles the intermediary play space of the world and the game, helps players to identify with the character, and envelops them in the game space. Players are participants who generate sound in their own physical space: the visual response happens on screen, but the sound happens in our own peripersonal space.

In addition to the many ways described in this chapter that sound contributes to immersion and identification, there are other ways in which interacting with sound encourages and facilitates identification with the game character, and immersion in the game space. I take these up in the next chapter.

3 Sound at the Borders: Enacting Game Sound

In the previous chapter, I discussed ways in which interactive sound situates the player into the space of the game, acting as an intermediary between the virtual and the real worlds and between the character and the player. The player's identification with the game character is facilitated by many different factors, but sound plays a significant role in the creation of space, in extending the body schema, and in delineating the boundary between self/character and other. This unique ability of interactive sound to situate the player in the game space, extend the virtual into the real, and drive identification with the character is further enhanced by the ways in which sound encourages and facilitates role-play in games.

Role-play is an important feature of many genres of video game because it is a way of testing out personalities, scenarios, and identities in a safe virtual space. Players often adopt a single avatar or character in a game and remain with that character throughout the game. In online massively multiplayer games, this adoption can mean weeks, months, or even years playing as the same character. The experience can be so compelling that the worlds sometimes collide, with role-played personalities augmenting or replacing real personalities in what performance artist Micha Cárdenas (2010) has called "transreal" identities:

A transreal identity is an identity which has components which span multiple realities, multiple realms of expression, and often this is perceived as a rapid shifting or a shimmering, as in the case of a mirage, between multiple conflicting readings. Millions of people today have identities which have significant components which span multiple levels of reality, including *Second Life* avatars and other virtual worlds. For many, such as the Otherkin or trans-species community, they consider these virtual identities to be their "true selves," more significant than their physical bodies. Yet the notion of transreal can be a way to subvert the very idea of a true self, if one's self contains multiple parts which have different truth values or different kinds of realness.

Role-playing allows players to explore these other potential, transreal identities and is an important part of game play, particularly the games that enable players to create an avatar that is their representation in the virtual world. Research has shown that virtual experience carries over into the real world (and vice versa) and that with today's perceptually convincing media, players' brains do not distinguish between real and virtual events. The longer that people stay in character in the virtual realm, the more likely that experience or character will spill over into their everyday lives (Blascovich and Bailenson 2011).

Although in some sense players may be able to transcend the body in the virtual space, they bring at least one element from the physical body—the voice. Sound thus can exist at a border between the virtual and the real, where it mediates and reconciles the two worlds. But the player's subjectivity and immersive experience can be both extended into these social interactions and disrupted by the player's sonic performances. And the body is not necessarily left behind in role-play gaming. By engaging the physical body, players may be better able to identify with the virtual body. Game sound can be a site of social and physical interaction where players actively perform their role and engage with the corporeality of sound.

In this chapter, I focus on three roles that sound plays on the borders between the character and the player—sound in music games, voice, and music as a mediator in an alternate-reality game. In each of these roles, sound helps to facilitate role-play, encourage engagement, and promote identification with the player's character.

Posing and Playing

Bodily engagement—particularly the mimicking of posture and gesture—is closely tied to the bidirectionality of emotion: emotions generate movements, postures, and gestures, and these movements, postures, and gestures generate emotions. If, for instance, we take on a slumped posture, we are more likely to feel sad and tired (Benyon, Höök, and Nigay 2010). We also tend to mimic the postures of those with whom we interact, encouraging empathy (the "chameleon effect"), even in the virtual world with virtual characters (Bailenson and Yee 2005). Ekman, Friesen, and Levenson (1990) have studied what they termed "emotional contagion," in which people mirror facial expressions that they see, eliciting the proper autonomic nervous system's response to the perceived emotion of others with whom they interact. In this way, a person's emotional state can be caught, and this has been shown to be true whether that person is real or virtual (Bailenson and Yee 2005). India Morrison and Tom Ziemke (2005) argue that such phenomena "can facilitate a user's identification with the character's 'body' as well as provide the groundwork for empathy." As with mirror neuron theory (discussed in the previous chapter), people have a strong tendency to translate what they witness into their own egocentric terms.

However, as was suggested in the description of audiovisual mirror neurons, in addition to bodily engagement with visualized characters and avatars, players have a similar bodily engagement with sound. They understand human-made sounds (including those of playing a musical instrument) in terms of their own experience of making similar sounds and movements. The mental re-creation of the sound causes a neuronal and motor-sensory response that mimics the performer/emitter, and thus players are able to interpret the emotional inflections through a mental re-creation of the action. People therefore give meaning to sound in terms of emulated actions or corporeal articulations (Leman 2008). Put differently, we mentally (and sometimes physically) imitate the expressiveness of the action behind the sound, based on our prior embodied experience of sound making (Cox 2001, 195).

The mental simulation of sound emitting and empathy with the performers of sound has been referred to as the mimetic hypothesis (Cox 2001), the ideomotor principle (Stöcker and Hoffmann 2004), kinematic empathy (Todd and McAngus 1995), bodily hearing (Mead 2003), anthropomorphic projection (Chadabe 2004), kinesthetic sympathy (Cage 1988, 95), and corporeal signification (Leman 2008), among others. Cone (1968, 21) uses the term "vicarious performance" to describe the physical act of listening—the mental performative activities that take place in which players may "covertly enact" the performance. In this way, "The 'completion' occurring in active perception is, strikingly, a performativity: hearing becomes doing. We turn the perceived into experience, and it becomes meaningful to us—touches us—through our bodily knowledge. The appearance of what is heard in the now is altered by that which it is (to us): an articulation co-construed by the musician's and the listener's performance—within our selves. The performance on stage is joined by a hidden performance in the listener's lived body: an enactment in listening" (Peters 2010, 87). In other words, we have a direct, embodied interaction with the sounds that we hear. John Cage's term for this is *kinesthetic sympathy* (1988, 95). Listeners have a kinesthetic sympathy with the creator of a sound source. In hearing music, for example, we mentally reenact the gestures of performers and are able to interpret their emotional inflections. The role of the body in understanding emotional content is critical to empathic engagement. Whether we are visually or sonically witnessing a person or game character, our embodied mimicry—our kinesthetic sympathy—of what we see and hear influences our ability to understand and empathize with that character. This suggests that when the ability of players to engage physically in this type of bodily understanding is restricted, their ability to engage with the content is altered.

When people play games, they often move the game controller and lean in the direction in which they want the character to move. The photos of video game players by Beate Geissler, Oliver Sann, Robbie Cooper, Shauna Frischkorn, and others illustrate the engagement of the physical body with the virtual space, and many people have watched children play a game in front of a television and seen them lean with racecars or characters, particularly in moments of stress, or jump the controller with the character (similarly, moviegoers may jump or lean in their seats when watching a film). Valentina Tanni (2001) describes the photographs of Geissler and Sann's *Shooter* (2000) (figure 3.1): "These physical movements replicate the moves taking place in the 3D space behind the screen, and the player's body becomes a conceptual extension of that of the avatar representing him or her, forging a bond between two universes which are only really distant in theoretical terms." Such behaviors increase the player's gaming experience even though these movements are not registered by the input devices (Newman 2002). This suggests that players may be moving mentally and kinesthetically sympathizing even when they are not physically moving.

Figure 3.1
Beate Geissler and Oliver Sann, *Shooter* (2000), photographs.
Source: Images from *Immersion Blog: Ideas, Bad Science and Art*, http://blog.robbiecooper .org/2012/01/22/beate-geissler-oliver-sann-shooter-2000.

But of particular importance to our discussion here is the ability of game controllers to encourage role-play and engagement with sound. With these control devices, sonic feedback is decoupled from the player's gesture: no matter how hard or how long players press that button, players still hear the same sonic feedback. Although, as discussed earlier, gestural input devices may be kinesonically incongruent due to the nature of unpredictability of the input, these controllers enable players to act out the emotion behind the action and thus heighten their emotion. Being able to act out the articulation in a sound allows players to understand the causality behind that sound and thus the emotion. We can therefore hypothesize that kinesonic congruence reinforces and enhances kinesthetic sympathy. Elsewhere, it has been shown that the gestures that we make influence the ways in which we hear the sound: by bouncing a knee to an unaccented beat, we impose beats onto the music and hear a distinct rhythm (Phillips-Silver and Trainor 2007). We move to the music, and the music moves us, which changes the way that we hear that music. If we extend this to all sound, then our movement influences the ways in which we hear and interpret that sound. Sitting still and listening are not perceptually the same experience as actively engaging with that sound.

Not only do gestural controllers help to facilitate identification with game characters through enabling sonic articulation, but gestural input devices encourage role-play. Kinesonic congruence, the idea of the player's gesture aligning with the sound that is heard, reinstates physically the causality that we mentally enact. With kinesonically congruent control over a game's sound, players have a sonic reaction that matches the kinematic action that their body is making, thus giving them the appearance that they are embedding their own expressiveness into the game space. Some research has shown that gestural controllers increase engagement and immersion in a game. Nadia Bianchi-Berthouze, Whan Woong Kim, and Darshak Patel (2007) tied engagement in music-based games like *Guitar Hero* (2005) to the extent of overall bodily motion that a player can make. They found that physical body movement by game players increased the players' levels of engagement and modified the ways in which they became engaged in the game (Bianchi-Berthouze, Kim, and Patel 2007, 111). By using gestural controllers (in this case, musical controllers like the guitar of *Guitar Hero*) and thus inducing body movement, players felt more of a sense of presence in the game world and had an altered affective state. The authors suggest that these effects are the result of the controller's ability to enable role-play: "The players appeared to quickly enter in the role suggested by the game, here, a musician, and started to perform task related motions that were not required by the game itself" (111). This connection might be strongest with music-based games, where kinesonically congruent sounds play a role in the actual gameplay.

Similarly, Sian Lindley, James Le Couteur, and Nadia Bianchi-Berthouze (2008) compared *Donkey Konga* (2003) bongos with a standard controller to determine how gesture affects social interaction, immersion, and engagement. They found that body movement increased the engagement level of players and encouraged greater social interaction: "By affording realistic movements, the bongos may have facilitated a willing suspension of disbelief during game play, and their flexibility may have promoted enjoyment by encouraging clapping and dancing. It has also been suggested that interfaces that require exertion promote engagement, as well as being a vehicle for social bonding" (Lindley, Le Couteur, and Bianchi-Berthouze 2008, 514). Such gestural interaction in games both encourages player gesture and demands it. Bart Simon (2009) suggests that gestural controllers lead to a gestural excess, where gestures are exaggerated beyond what is necessary to play the game. In other words, gestural interaction encourages the player to act out, perform, and exaggerate the game's character. As Allison Sall and Rebecca E. Grinter (2007, 220) describe, "Participants often used the term 'performance' to describe their physical gaming experiences. The presence of others cast in the role as spectators became necessary to transform the player into the performer."

The role of the spectator is particularly important for the player in music-based games. By its nature, the auditory has spectatorship and performance elements (Godøy, Haga, and Jensenius 2006), and games built around sound often encourage performative—and competitive—play. There have been extensive competitive events around *Dance Dance Revolution (DDR)* (1998), for instance, which at the peak of *DDR*'s popularity in Asia were large enough to fill stadiums of spectators. *Dance Dance Revolution* and *Guitar Hero* games both make use of spectators inside the game as well as externally, having an in-game crowd that boos or cheers the player depending on the player's success (Smith 2004, 70). The competitions highlight the performative nature of the genre: *Rock Band, Guitar Hero*, and *Dance Dance Revolution* have all had many public competitions in highly public places. Electronic Arts' 2007 Be the One event was held in London's Trafalgar Square, and a *Dance Dance Revolution* contest was held in Sony Metreon in San Francisco.

Johanna Höysniemi's (2006) survey of *Dance Dance Revolution* found that although most players of the game followed the game's rules and goals, others focused on the performative aspects of play, "whose main reason for playing is to dance expressively either with or without having practiced a routine. . . . One of the main purposes of freestyling is to entertain and impress the audience. And many freestylers commented that they want to make the crowd laugh, and so made up all sorts of funny moves and crazy stunts." Players described the importance of the audience as critical to their enjoyment of the game, and often they were drawn to play the game because of their own experience as a spectator. Indeed, the success of these types of games is in allowing the player to perform. In the case of *DDR*, as Alexander Chan (2004, 5) describes, "The raised metal platform with embedded arrows was an enormous departure from other game interfaces because it implied a performative aspect to play; it required full movement and rhythmic motion which was often interesting to watch. Furthermore, the platform was often called a 'dancing stage,' and being raised slightly above ground level, the game interface paralleled a theater stage on which performances take place." The large platform that comes with the arcade version of the game requires considerable floor space in an arcade encourages spectatorship, in the sense that far more spectators can watch a player on the *DDR* stage than with older more traditional upright arcade games (Andrews 2007).

Dance Dance Revolution was the most successful of the music-based games until the arrival of *Guitar Hero*. As noted above, music-based games like *DDR* and *Guitar Hero* by their nature imply spectatorship: they are "theatrical by design" (Miller 2009, 401). Kiri Miller (2009, 419) argues that *Guitar Hero* is designed to attract real spectators who watch the avatar band on-screen (which due to the nature of gameplay is lost on the player, who is too busy following musical notes) and that "once an audience has gathered, many players feel pressure to put on more of a show themselves." Spectatorship is implied by the gameplay itself: the game takes place in a variety of venues that place the player/avatar on stage, and a virtual crowd boos and cheers the players, even when they have no real spectators. There are also many YouTube videos of performers that ensure that players can demonstrate their virtuosity even when no live audience exists during play. Indeed, such games "offer little freedom of expression apart from the prerogative to perform while playing" (Pichlmair and Kayali 2007, 426). Such spectatorship promotes role-play: without an audience, there is less incentive for players to embellish their play and act out the "hero" role of star performer.

Ethnomusicologist Kiri Miller (2012, 16) suggests that *Guitar Hero* is a form of *playing between*—"playing in the gap between virtual and actual performance." Although the playing may largely be mimicry, the performance of play is often real. The game provides performative "hints" that encourage the performative activities that take place during play: it is less a guitar simulator and more a "rock performance simulator" (121). Thus the guitar controller offers a new means of musical performance. Miller (2009, 402) describes six performances that occur simultaneously in a single video performance of a *Guitar Hero* track—"the edited video performance designed for YouTube, the living room performance, the avatar's performance on-screen, the human performances in a motion-capture studio that provided the physical model for the avatar's performance, a studio band's cover performance of the song 'YYZ' (the game development company wasn't able to license the Rush recording), and Rush's studio recording of the song"—and argues that player Freddie Wong's performance in front of his friends in his living room is "closest to the rock ideal of an authentic live show" and the most theatrical in its parody of rock authenticity.

Games and virtual worlds have helped lead to new forms of sonic performance that question notions of liveness and authenticity. Players engage in a variety of simultaneous performative activities that become a form of play. Performance in itself, in other words, becomes a playful part of the game experience. Through promoting performance, these types of games strengthen the role-play aspects of the game, encouraging a kind of rock performative excess and rewarding the player for these gestures (raising the guitar to enter Star Mode and so on). Shy players may in this way mimic the kinds of performative activities that they witness popular stars enact and through this physical mimicry may be able to live that performance mentally.

Although other types of gestural game controllers may mimic real-life objects (such as steering wheels and guns), the guitar and dance stage resemble existing performative practices. The *Guitar Hero* series uses a dedicated guitar-shaped controller that mimics some of the aspects of playing guitar (see below), thus allowing the player to take on the character's performative role in the game. As described above, in hearing music players mentally reenact the gestures of the performer, and with music-based game instruments this becomes a reenactment of emotional content in music through performance. Rolf Inge Godøy, Egil Haga, and Alexander Refsum Jensenius (2006, 256) suggest that the imitative actions involved with air guitar can change the ways in which players relate to music and also illustrate how they perceive emotional content in music: "Both musicians and non-musicians can often be seen making sound-producing gestures in the air without touching any real instruments. Such 'air playing' can be regarded as an expression of how people perceive and imagine music, and studying the relationships between these gestures and sound might contribute to our knowledge of how gestures help structure our experience of music."

Current musical game controllers are limited in their expressive abilities: players can lift the neck of the guitar in *Guitar Hero* and to some extent employ the whammy bar, but they cannot employ the nuances of vibrato, palm mutes, bending, slides, or harmonics, and the strength of the button presses makes no difference to the notes produced. The loss of the articulatory, expressive elements of gesture arguably lessens the emotional involvement of the player. The imperfections and inconsistencies that signify liveness and personality in an acoustic recording are penalized in the game because they are interpreted as errors. If the player could add these expressive elements through controller design, then they would also increase their involvement in the game, their feeling of connectivity to the original performance, and even their empathy for the character.

Part of the widespread appeal of the *Guitar Hero* series[1] involves the different types of gestures that occur in playing a musical instrument. Godøy, Haga, and Jensenius (2006) distinguish among sound-producing gestures (those that relate to producing the actual sounds), sound-accompanying gestures (such as dancing and sound-tracing), and amodal affective or emotive gestures (those associated with global sensations, like effort and velocity).[2] They later simplified this distinction to just sound-producing and sound-accompanying gestures (Godøy 2010, 110). With *Guitar Hero*, players can pose with the controller (the guitar) as though it were a real guitar, swinging it and raising the neck in emphasis. These actions do not affect the sound but influence how the controller makes players feel as if they are "really playing." in fact, the game rewards these types of actions with "star power" if they occur at the correct time. Nevertheless, the actual sound-producing gestures of the guitar controller—pushing buttons and strumming a plastic knob—are nothing like those used with real guitars, which require fingering frets and strumming strings. The embodied listening to and interacting with music in these games encourage a form of "collaborative performance: the players and their audiences join the game designers and recorded musicians in stitching musical sound and performing body back together" (Miller 2009, 424).

People enjoy playing *Guitar Hero* because it allows for the emotional mimicry of playing a real guitar: it is the emotive expressiveness of sound accompanying that drives the performative aspects of the play. As the designer Rob Kay (in Miller 2009, 413) describes,

I think we're always kind of keen to get people doing that move [star power] because . . . it's really obvious that once you give people this cue that their physical performance has got something to do with the game, even though that's the *only* physical performance that has something to do with the game—you're suddenly in the mind-set and thinking about all of those rock-star moves that you see people do, and then they will just jump around and do the rest. And it seems that that's often the case with these kinds of things. You don't have to go the whole way. You just need to give people a beginning—a shove in the right direction, and they'll do everything else.

We might also return to our conception of the distinction between evoking sounds, shaping sounds, and creating sounds to understand the distinctions in musical game performance. In games like *Guitar Hero*, the player is able to evoke sounds and, to a limited extent, even shape those sounds (through the use of the whammy bar) but is unable to create sounds. Through the performative aspects of play, however, players are afforded the illusion that they have some creative control. By enabling and encouraging performance through physically enacting the emotional content in the music, the game's creators allow the player to take on the identity of the player-character. In addition to embodied musical listening, role-play in games is encouraged by another interaction with sound: the voice.

Voice and Role-Play in Games

The theory of embodied listening described above, relating to the ways in which we bodily or mentally mimic sound effects and music, carries over to voice as well: we hear the voices of others by experiencing those voices in terms of our own bodily postures and expressions. And just as we mimic the physical postures and facial expressions of those with whom we inter-act in order to empathize, we likewise mimic voice. During a conversation, speakers even adjust their own voices in a way that brings about a conver-gence in speaking pace, phonetic pronunciation, frequency range, and so on (see Pardo 2006). Voice therefore can help players to identify with and empathize with the game characters.

The use of character voices in games varies considerably in approach. Some games are completely unvoiced and rely on written on-screen text, which was common in the days prior to adequate sampling rates that could reproduce voice at a reasonable quality. Some games jump back and forth between text and voice, depending on whether the material is interactive or presented in a cinematic sequence. The *Professor Layton* (2007) series on the Nintendo DS, for instance, oscillates between using voice and using text. In some places, players control the play-back speed of the dialog, in others the player has no control over voice playback, and in still others there is no voice at all. In part, this is likely due to the amount of narrative and dialog in the game: it is faster for the player to read the text than it is to wait for the characters to speak through their script.

The *Professor Layton* series of games had different voice actors for the North American and European versions for the child character, Luke, illustrating that getting the right voice for the character differs depending on audience. The American version had an affected (and toned down) British accent (voiced by American actor Lani Minella in the U.S. version), and the European version had a cockney accent (voiced by British actor Maria Darling). Online debates reveal that most Britons dislike the American version because the accent is not heard as authentic enough. Americans, on the other hand, feel that the cockney version is "too British."[3]

US Luke is insulting to the British. He sounds like an American trying to do a posh accent, but failing. We Brits sound NOTHING like that. (LegalKlaura)[4]

I have to admit, I do prefer the UK version of Luke because, after all, Luke is English. His accent, despite rather annoying at times, is an actual, traditional commoner cockney accent—kind of like Oliver Twist with more fire. (kittykatkrisa)[5]

What is particularly interesting in the debate over Luke's accent, is not so much which voice the players preferred, but that the voice acting makes enough of a difference that many Brits said they could not listen to the American version. Although to some extent the players have been conditioned to the voice that they heard during several episodes of the game, they felt strongly that the wrong voice would make the game unplayable.

Some other games have relied primarily on text for dialog but have added a few exclamatory voice clips to give the characters a hint of voice. Nintendo's *Super Mario Bros.* (1985) series uses this technique. In these games, Mario usually has a few words in his vocal repertoire (such as "Luigi!" and "Mama Mia") but for the most part remains voiceless.[6] One reason for this lack of voicing of the character could be that the Mario franchise became popular before vocals were feasible for characters. When Nintendo has provided the series' characters with voices, the critical response has been negative: "Nintendo and NST also go just one notch further in over-the-top presentation by making Mario and Donkey Kong overly chatty. Whether it's the direction Nintendo's taking with its key characters or simply the development team showing off its audio compression techniques, it's an element that just doesn't need to be there. It's not overly distracting to hear all this chatter, but it would just fit the characters more if Nintendo just kept the dialogue simple" (Harris 2004). Long-running franchises risk alienating fans who have played the unvoiced versions of the games and have already preimagined the voice in their own mind. As with the *Professor Layton* voice changes between continents, when players are accustomed to a particular voice (whether real or imagined), any other voice seems unnatural.

The announcement of a new *Legend of Zelda* series game (another long-running Nintendo franchise), *Skyward Sword*, at the Electronic Entertainment Expo (E3) 2010, for instance, led to a debate about character voice on the IGN boards:

One thing I really hope, is that they don't do voice acting and if they have to it shouldn't be for [main character] Link at all. I never what to hear his voice, because of the pure epicness of the character no voice would really fit. (EdElric666)

Awesome!!! The only thing I can ask for now to say this game will be perfect: Voices! Let the characters speak!! Let us hear them! (JG149)

I have to be honest with you guys. Don't expect Nintendo to ever do voice acting in a *Zelda* game. Why? Because Link is not Link in the game. Nintendo gives their players the ability to name their character. Sure we always identify with the name Link, but the non-playable in-game characters always refer to Link by the name you give them. Thus making voiced dialog in the game impossible. Unless Nintendo eliminates the "name your own character" feature of the game, which I honestly doubt will ever happen because it's been a constant feature of the game since the original, you should expect to continue to read through Zelda games. (Great_Scott)

Link doesn't have to talk for voice acting. It would be fine if he always remained a mute. But the [nonplaying characters] definitely require voice. After a while, mixing written dialogue with grunts and yelps just doesn't cut it anymore. It was a travesty that none of the characters in *Twilight Princess* had a voice. I bet they could have found a really cool one for Midna, too. (SlitHaHs)

. . . On a side note, what's with people's obsession with voice acting? If people made noises instead of text boxes, then how does that improve gameplay? Is that [going to] give us fresh fun new puzzles? Is it going to give us harder bosses? Is it going to give us a more original story? Is it going to change the structure of the game? Voice-acting has no value in terms of gameplay, which is where our demands should be aimed at. Sure, it gets rid of text boxes, but reading wasn't really that big of a problem. People are caught in the idea that voice-acting is the "way of the future" and that because every other game has it, that Zelda is behind. That's just stupid. Voice-acting is for cut scenes and making video games feel like movies, I'm not focused with making Zelda a movie. I want Zelda to be a good "video game." Just ask yourself, if [*Skyward Sword*] had voice-acting, are you going to have any more "fun"? I just think people need to prioritize demands a bit better (I can say the same for graphics). (Crescent_Soul)

. . . I don't think it should happen any differently than the voice acting in *Twilight Princess*. Imagining the voices in the game is one of the things that makes each Zelda game unique to each person that plays it, it gives you the fond memories that last, plus, you don't ever have to worry about horrible voice actors. (NortheastBat)[7]

Although some players would like to hear the character voiced, most players felt that they had already constructed their own voice (and likely, by extension, a persona) for Link. By voicing the character, the player's entire conception of that character may change.

The *Zelda* debate also raises the issue of the role played by character customization and the voicing of the player-character. When players play games, they may have a character preselected for them, or they may be able to customize their character. Many players feel strongly that voice should be left out of games in which we can customize an avatar. *Little Big Planet* (2008), for example, allows players to customize their characters (sackfolk) completely and leaves out voice altogether. The sound designer Kenny Young (2009) described his approach as follows:

I think the sound design decision not to litter the Sackfolk with inane voice samples contributed significantly towards [winning an] award. This is something which some sound designers find hard to resist—there are a couple of 3rd party trailers and adverts out there where some arse has added chipmunk voices to the Sackfolk to make them sound comic and cute. . . . not having a voice allows players to feel that their Sackboy, which they lovingly dress, customise and emote with, is theirs. This sense of ownership would be hard to achieve if the character you were controlling had a mind of their own, voice and language being the most personal way of communication and expression. Which is why when Sackboy does speak, it is with the voice of his player, his lips moving to match those of his puppet master.

Young's point here is that the role of the sound team is, in part, to make players engage more fully with their character. Young argues that the best approach to player-character dialog is to eliminate the voice of the player's character because players have already read the words on-screen and thus "heard" the voice in their heads. If players are to become a character, shouldn't they sound like the players or at least how they imagine characters should sound?

Research into avatar customization has been shown that, when given the choice, people tend to select avatars that are similar in appearance to themselves with a few enhancements (Messinger, Ge, Stroulia, Lyons, Smirnov, and Bone 2008). This suggests that to create a character with whom they can identify, players may similarly wish to sound like their character. Some games that have allowed for a high degree of character customization have also allowed for some degree of customization of voice.

Figure 3.2
Voice customization in *MySims* (2007).
Source: Image from the *MySims Wiki*, http://mysims.wikia.com/wiki/Sims.

MySims (2007) and *The Sims 3* (2009), for instance, have voice "sliders" that allow players to adjust the vocal tone of their avatar (figure 3.2). In *Spore* (2008), another game that relies heavily on player customization, players can select a mouth size and shape for their creature, and those selections adjust the character's volume, pitch, and language. The overall timbre or sound of the creature can be changed by using multiple mouths. The voice adjusts throughout the game, however, as the character evolves and develops language. Although the options have to be limited to make the development of the game manageable, the scope of such customization is changing quickly as the technology advances, and players can expect to see more voice customization options in games in the future.

In some recent popular massively multiplayer online (MMO) games like *Dragon Age: Origins* (2009) and *Skyrim* (2011), the player-character does not have a speaking voice but has various grunts, shouts, and other exclamations that were recorded for the many different races and sexes of the potential player-characters. Creator BioWare's explanation for the absence of player-character voice is that the amount of customization available makes prerecording voice actors unfeasible (Totilo 2008). Players have discussed the absence of the character voice in these games in discussion forums online, with many players concluding that having a voice in these particular games would be detrimental to the immersive effects of roleplaying of the main character. One advantage of leaving the voice out is that an absent voice can never be the "wrong" voice, according to the player:

I'm happy I'm not going to have a voice in *Skyrim*, or else it would totally kill the immersion. And as I saw higher in the posts, sometime they give your character a funny voice. Any one of you played *Red Dead Redemption*? Remember Jack's voice? I swear I was going to dropkick [developers] Rockstar in the FACE after that one. (Bobbyleez)[8]

I think it would ruin the RPG experience if they gave me a voice like I can hear my character talking during dialogue. It's an RPG game that lets you be whoever you want. Having a defined voice would ruin it for me. It's nice for games like *Deus Ex* or *Mass Effect* but definitely not for *Skyrim*. In *Deus Ex* and *Mass Effect*, you play as a defined character, the one with a specific background story and personality (maybe not for [*Mass Effect*] but you only have two to three options anyway). In *Skyrim*, you are a prisoner being led to execution and you happen to be a dragonborn for some reason. You'll find out why you're dragonborn through the quest I guess but your character's background story, personality and appearances is your choice. (hexperiment)[9]

Also to anyone who complained that your character didn't have a voice in *Dragon Age Origins* and sees that as a flaw in that type of dialogue system. Who gives a shit!? Really? You don't have the sliver of imagination it takes to picture how your character might sound like? It's like complaining about books because they aren't being read out aloud to you. Do you think "Hmm, it's strange. In this book they keep talking but yet I can't hear anyone talking. Then you close the book and put your ear down to it real close. Still not a sound coming out of the book. What a crap book. They're all mutes! Now I don't even know how they sound like [in real life]!" I for one loved the fact that he didn't have a voice because that made him have the voice that I wanted him to have. Not some actor that might not sound anything like I pictured my character to sound. (OkTank)[10]

Well, a part of it has to do with immersion. In *Half-Life*, you play as a silent protagonist, Gordon Freeman. People around you will talk to you but you never talk back. But it doesn't feel awkward at all. The world is interacting with you and presenting itself. It really feels like I am that character and I can imagine whatever dialogue that could've been there. You feel like you're Gordon Freeman even though he probably doesn't have your thoughts. No voice makes it more immersive. If Gordon Freeman began talking, sure we'll get more personality out of him but you'll be put into third-person perspective, even if you're looking through the eyes of Gordon Freeman. . . . Even if I can choose which line to say, it's still not "me."

Before you know it, your character would start having his own opinions. That eliminates the whole theory of "you can be whatever, whoever you want." The only time I want a character to have its own voice is if the character is well-defined and like-able, and his or her personality would play a very strong role in the story. (hexperiment)[11]

I guess when I play a character with a voice, I just imagine it as my voice and it never really creates less immersion, except when it doesn't say what I want it to say. (Tusck)[12]

In these instances, most of the players feel that by not being given a voice by the game's designers, they can provide that voice themselves, imparting their own voice to their character, giving the character their own personality, and increasing their role-playing ability and the immersive aspects of the game. If the voice in players' heads that is ascribed to the character (perhaps the player's voice or perhaps an affected voice) is not congruent with the voice that players hear, they may lose their connection to the character. Given the inability of game producers to create a multitude of voices for such massive games, many players (and apparently developers) agree that no voice is better than the wrong voice.

In the more recent incarnation of the game, *Dragon Age 2* (2011), players do not have customizable avatars but instead select a specific predefined male or female character, Hawke, who has been provided with a voice. *Dragon Age 2* allows players to use a dialog wheel to select what they would like to say based on three personality types (diplomatic/helpful, humorous/charming, and aggressive/direct). The selected text decides what Hawke says and alters the character's tone and personality during the game.[13] Hawke learns the tone to use in future scenes by the player's repeated selection of that personality type, thus altering the playback of the rest of the game. Some players like this system of defining not just voice but personality (Young 2010b):

Dragon Age . . . uses a different system whereby you are presented with several verbatim options for what your character could say and, then, as soon as you click on one of these phrases it is as if your character has already said it and you immediately hear and see the other party's response. This works beautifully for several reasons: Having read all the options, considered whether it fits with the character you have established and any potential outcomes, there is no need for you to hear your character speak this information out loud again (a trap fallen in to by earlier games, such as Ion Storm's *Deus Ex*) because you've already just "heard" it in your head when reading it. And so, the act of clicking replaces the act of speaking . . . by not hearing a prescribed character voice which takes them out of the experience the player is empowered to fully inhabit their character.

Nevertheless, in many conversations, the amount of text that is required to display the conversation options in the dialog wheel meant that what was written was often shortened and what was voiced was not what was written:

This [inability to map text to actual spoken dialog] . . . detracts from character interaction and immersion because:

You don't really know what your character is going to say or do. This rips the characterization away from the player and further distances them from the role-playing aspect of an RPG.

You don't know how your character will react and this can result in you needing to restart saves in order to "fix" an action that you didn't actually intend to do. Yes, I realize they want to put a "mood" indicator next to the words, but that doesn't fix what is flawed with the conversation wheel. The *Mass Effect* team's solution made it just as easy for you to know what response was "good" or "bad" and it does not help. One of the more startling parts of *ME2* was when you flat-out murder someone when you thought you were simply going to incapacitate him. It's very jarring when your righteous and good character suddenly just butchers a character because you didn't know what their REAL response would be in comparison with the 2-word paraphrase.

Worst of all, it severely limits the number of responses that can even be displayed on-screen at a time, thus limiting the possible ways that you even can respond. How many *Dragon Age* conversations had half a dozen to a dozen responses? Quite a few, especially the more important conversations. You never felt like you were just flipping a coin between 1–3 responses that were "good bad and neutral" you were actually steering your character's personality via many responses that, while similar, were subtly (or blatantly) different. To summarize, the conversation wheel

• limits the number of possible responses
• detracts from character immersion
• limits on-screen words and results in paraphrasing that never fully reflects what your character will say or do, resulting in you throwing the dice whenever you pick a response.
• limits the emotional range of the main character.[14]

And most important, the idea of immersion and identity breaks down again:

It also changes the game from first to third person, not a plus for many of us. There's no pretending that you're Hawke if you don't know what you're going to say or do until it happens. (errant_knight)[15]

The biggest complaint I have is the implementation of the *Mass Effect* dialogue system. One of the main reasons I loved *Dragon Age: Origins* was the intricate moments you would have while in dialogue with your party members. I was astounded by the sheer amount of choices in any given dialogue sequence I had with my companions and NPCs. In *Dragon Age Origins* there were so many ways I could respond to something that I never felt that my character was shoehorned into any defined path of how his personality ultimately was. In *Dragon Age 2* the choices you're given in dialogues ultimately boil down to: Tell me more about something, understanding answer, HI-LA-RIOUS rogue-ish answer or aggressive answer. Congratulations BioWare. You solved the problem with *Mass Effect*'s dialogue system being binary by making it TERNARY. I don't want ternary. I want fucking octodecimal or something to the degree that *Dragon Age* origins had in its dialogue tree. Moreover, how the hell am I able to discern what my character is going to say when all I'm presented with are dialogue phrases that are so short they make tweets seem like wall of texts. I have NO idea how Hawke is gonna word a response from these phrases. Countless times have I picked a choice because of how I thought Hawke's response was going to be from those 3 fucking words you have to go on, only to have Hawke say something completely different.

One example: Upon meeting Isabella and talking to her in the tavern she told me about her problems. Knowing I talked to a swashbuckling rogue I chose the "Hi-LA-RIOUS" choice. The phrase linked to that was something like "Need help?" which I thought was going to play out kinda flirty and smarmy *wink *wink and all that. Instead Hawke says "Can't Anyone Fix Their Lives around Here!?" in an annoyed tone . . . not what I wanted to say buddy, not at all. . . . There's countless examples of this throughout the game. As far as having freedom to control how your character acts I hate this system with a passion. All because of bloody *Mass Effect* and how cinematic it is. BTW the word cinematic inherently means "less immersive" to me. It means that dialogues are going to be really dynamic like in the mooooooovies. It's like you're watching a mooooovie. I don't wanna watch a movie, I don't wanna feel like I'm in a freakin' movie. I wanna feel like I'm in a world where I'm writing the script and I'm defining my character. Not watching a movie where I'm the lead actor. As Brian Lehey of *Weekend Confirmed* once said, "In *Mass Effect* you're Commander Shephard: in *Dragon Age Origins* the grey warden is YOU." Very important distinction. (OkTank)[16]

This latter point is revealing: rather than playing a character, some players want to feel as though they themselves are in the game. They do not want to role-play someone else but want to role-play the game with themselves as the character. They view voice as being critical to this ability to role-play.

Many players feel so strongly about the dialog wheel that a dialog wheel spoofs became an Internet meme (figure 3.3). One person built a Mass Effect Dialogue Wheel Generator that lets players type phrases and choose wheel settings to superimpose over a background image.[17]

Because players influence the outcome of the game through what they say, the dialog wheel places a high degree of importance and focus on the voice and dialog. Players cannot simply skip dialog sessions (as is common

Figure 3.3
Dialog wheel spoof poster. It reads "I'm in love / I'm happy / I'm weak / I'm confused / I'm frustrated / I'm angry" in the dialog wheel, with the explanation "Your character development in a nutshell."
Source: Image from *Tellyr of Tales*, http://tellyroftales.com/?p=758.

in many games, particularly on repeat play-throughs) but must actively engage in the dialog to achieve the outcome that they desire. Moreover, the dialog is not relevant only on the first play-through. Players may opt to replay the game selecting a different personality type/dialog for the character. Nevertheless, the current limitations of the system (particularly not being able to read exactly what the character will say) mean that players lose that empathic connection to their character because the character is not responding as they expected.

Repeated throughout the players' explanations quoted above about whether they like or dislike the voice systems of games is the idea of immersion and identity. The idea that players might not be able to become involved with a character with the "wrong" voice fails to take into account the power of role-play in games. We mimic voice all the time in recounting a conversation, for instance, and it may be that we are comfortable taking on another's voice providing that it is believable and that we understand the emotions behind that voice. The character does not look like the players or have their hands when they reach out into the game (when playing in first person), so perhaps players are not more concerned about the voice matching expectations than the visual identification of the character.

Some players see the voice as so intimately personal that they have a hard time mentally adopting the character's identity when provided with a voice. After all, if a character (such as a large Orc) does not physically resemble the player, then the two voices would not be similar anyway. With voices, the importance of kinesthetic sympathy might be heightened. In games, characters are animated, and players are limited in terms of the gestures that they can make in the game. The character is always animated: it is not visually a real person or real creature. The voice, on the other hand, is a real voice. As with the synchresis of sound effects altering the visual image of on-screen objects, voice alters the visual character of the avatar. That synchresis that takes place brings a realism to the character through players' own empathic, embodied listening experience. Taking on the character's voice causes players to re-create its emotional state and posture through their empathic connection to real voices. If players fail to identify with the character and make that mental mimicry, then they may not be able to feel connected to their own character. Sound, in other words, is critical to an empathic engagement with the character.

So the wrong voice can break with immersion and identification in games. But what happens when players use their own voice in a game, as in online multiplayer games with voice chat? Multiplayer games, in which players can play with or against each other, are one of the fastest areas of growth in the game industry. *World of Warcraft* (2004) has over 10 million subscribers (Holisky 2011). But along with massively multiplayer games (MMOs) where thousands of players may be online playing a game at any time, there are many other forms of multiplayer games. Many games today are released with an online element that enables smaller groups of players to play with or against each other over a server (such as Xbox Live or the PlayStation Network), and most console games released today have multiplayer options that allow several players to play in the same physical space.

The structure of many multiplayer games requires players to interact with each other, since players are dependent on cooperation to progress through various stages of the game (Caplan, Williams, and Yee 2009, 1313). Most players rate this social aspect of gaming as the most important factor for playing, with a majority continuing to communicate with other players outside of the game on message boards to schedule playing, exchange advice, and socialize. Constance A. Steinkuehler (2006, 50) argues that MMO gaming is itself a form of discourse, with "fuzzy boundaries that expand with continued play. What is at first confined to the in-game space alone (between log-on and log-off) soon spills over into the virtual world beyond it (e.g., Web sites, chat rooms, e-mail) and even life off screen (e.g., telephone calls, face-to-face meetings)." Various forms of communication in and out of the game are therefore a critical element of success in the games. In-game, this communication is typically done through voice-over Internet protocol (voice-over IP, or VoIP).

Voice-over IP in multiplayer gaming has been commonplace since the rise of broadband Internet in the early 2000s. The current most popular software applications that enable VoIP are Xbox Live, Roger Wilco, Teamspeak, and Ventrilo (referred to by players as "Vent"). Voice channels are configured like a telephone conference call, with members of a specific team connected to each other. In this way, teams of players can conspire and discuss gameplay tactics, share information, and talk about the game or about things external to the game to members of their team without sharing this with others in the game world.

Initially, networked games used text messaging rather than voice. Text requires players to type, however, and they cannot type and play simultaneously. With voice, the player's hands are freed, and therefore coordination between players can happen more quickly, a critical point in ensuring the team's success in some games. Text takes the body out of the discourse: players cannot hear its emotionality and cannot empathize in the ways described above. Voice can also pick up the nuances of speaking that can be misinterpreted in text. Emotion, personality, and mood can be more easily conveyed through voice (Wadley, Gibbs, and Benda 2005). In some sense, the use of VoIP ends the feeling of mediated interaction among players because the voices and dialog of all characters in the game-space are real and the responses instant and apropos.

Another important effect of voice communication is the sense of social presence that is provided by voices in the game space (Duchenaut, Yee, Nickell, and Moore 2006). When playing, the player is surrounded by conversations, and even if the player is not at that moment communicating, voice creates a sense of being in a public social space. Evidence suggests that VoIP creates an increased sense of presence through increased engagement and energy (Halloran, Rogers, and Fitzpatrick 2003). In their work on the online version of *Counter-Strike* (1999), Talmadge Wright, Eric Boria, and Paul Breidenbach (2002) argue that "The meaning of playing

Counter-Strike is not merely embodied in the graphics or even the violent gameplay, but in the social mediations that go on between players through their talk with each other and by their performance within the game. Participants, then, actively create the meaning of the game through their virtual talk and behavior." Much of the interaction between players takes place through voice, and much of the atmosphere of the game is created through voice.

There are some problems with VoIP as a communication device for games, however. Some players feel that voice adds little to genres of games that are not based on collaborative play and that the use of the voice channel in those cases degenerates primarily into trash talking other players (Wadley, Gibbs, and Benda 2007). The tasks that take place in games are varied: some of my relatives meet online once a week to compete in racing cars, for instance, which has different goals than a collaborative MMO role-playing game like *World of Warcraft*, where players need to cooperate with each other and share information. Moreover, the practice of grinding in role-playing games means that players have time to engage in social conversation as a means of reducing the boredom of these parts of large multiplayer games.[18] In other genres of games, voice is not as helpful, however, which helps explain why voice is sometimes maligned. A well-known machinima video known as *Leeroy Jenkins* (2005) highlights one of the issues with voice communication: it has a temporal limitation, whereas the text-based method allowed players to scroll back and view the conversation. In the video, player Leeroy Jenkins steps away from his computer during a game of *World of Warcraft*. Members of his players' guild, "Pals for Life," are meanwhile discussing strategy and projected death count before heading into battle. Leeroy, coming back to his computer, runs straight into battle, ruining the strategy and leading to the death of all the players' characters. It was later revealed that the situation had been preplanned, but the point is that when communication is asynchronous, voice is less effective than text.

Furthermore, when many characters are on-screen, players might not know the person to whom they are speaking (virtually or in reality), and it may be difficult to match the voice to the avatar (Wadley, Gibbs, and Benda 2005). Likewise, when too many players are on-screen or there are some particularly vocal players, there may be a lot of simultaneous conversations, which are harder to track in voice than with text. Further confusion can be caused by an inability to know if one's own voice has been heard. With a lack of visual cues, not hearing a response to a question or comment means that players are sometimes left repeating themselves

(Gibbs, Hew, and Wadley 2004). Referring to existing systems, Dolby Laboratories notes that "there are lots of stress-points in the system. You have players using mono headsets and low quality microphones, the codecs aren't very good, and there is also the issue of people who have their microphone settings too low or too high. The results even on relatively controlled services such as Xbox Live are clipping, distortion, and microphones picking up echoes: essentially an audio mush" (Arnold and Langsman 2009, 18).[19]

A significant criticism of VoIP in games is that the illusionary aspects of role-play gaming can be destroyed by many different factors. One of the most significant sound problems with VoIP are the "ludic leakages" that occur: game sounds seep into the real world space (and vice versa) and disrupt the immersion (Pearce, Boellstorff, and Nardi 2009, 177). Family members in a nearby room can overhear a private conversation, for instance, which means adjusting vocal volume and often regulating language. When players can freely swear, share secrets with strangers, and otherwise engage in behavior not appreciated by other members of their physical family, they may be embarrassed by talking out loud and prefer text. This means that the player must censor themselves, which potentially reduces immersion. Likewise, the voices of family members who speak to or near the player may be picked up by the microphone and enter the game (Wadley, Gibbs, and Benda 2007). A player's mother's vacuuming, for instance, may disturb other players. Some of the talk that occurs is out-of-character (such as asking for technical or help related to game mechanics), which means that role-play can be further dissolved by voice.

The role-playing aspects of the game also can break down when speakers are clearly of different age, gender, or ethnicity than that of their character. The social baggage of gender, ethnicity, race, and age always comes into play when players use their own voice (see, e.g., Debevec and Iyer 1986). As discussed above, a voice that fails to live up to the player's expectations can destroy the illusion. Although equipment has been designed to disguise the voice through the microphone (Download 3K 2006), the unnaturalness of the voice through modulators can itself be a distraction, and many players prefer not to tweak their voice (Wadley, Gibbs, and Benda 2007). This inability to disguise the self destroys the suspension of disbelief, as Richard Bartle (2003) explains:

If you introduce reality into a virtual world, it's no longer a virtual world: it's just an adjunct to the real world. It ceases to be a place, and reverts to being a medium. Immersion is enhanced by closeness to reality, but thwarted by isomorphism with it: the act of will required to suspend disbelief is what sustains a player's drive to be, but it disappears when there is no disbelief required. . . . Adding reality to a virtual world robs it of what makes it compelling—it takes away that which is different between virtual worlds and the real world: the fact that they are not the real world.

Bartle's criticism of voice in some ways echoes that of the players above in describing how the wrong voice makes it difficult to adopt the player-character. As noted above, although players have little difficulty adapting to fantasy if the visual character does not match their expectations, when the voice—an artifact from the real world—is involved, the stakes become higher, and that intrusion of reality into the virtual can greatly affect immersion.

But some aspects of speaking can increase players' identification with the character and thus increase immersion. By physically taking on the vocal traits of the character, they are acting out the character in the game space but are also (probably) at the same time altering their own physical body posture to enact that role. Even though other players cannot see the gesture, gestures are still used in speech generation (Kendon 1997). Gestures are closely related to speech production. Indeed, some evidence suggests that they are connected into a single system in the brain (Barbieri, Buonocore, Dalla Volta, and Gentilucci 2009). Since we know that we can feel emotion through emotional contagion and the mirroring of other people's gestures and postures (Ekman, Friesen, and Levenson 1990), it seems likely that by affecting a voice and taking on the posture and gesture of a character, players may further enhance a connection to the character in a way that would not occur with typing text or selecting options from a dialog wheel. Speaking might activate gesture, and this activates the same mirror neuronal response that was described earlier: affecting a character's voice involves affecting the posture and movement of that voice and thus the emotion behind it. Through a performance of the character's voice, players may increase their engagement and involvement in the game.

Alternate-Reality Games

Another aspect of gameplay in which players take on the role of the character while interacting with sound is alternate-reality games (ARGs) (or mixed-reality games). Alternate-reality games are a form of pervasive game that exist primarily offline in the real world. Instead of adopting an avatar, players are the game characters in a kind of variation of live-action role-playing. The game story is delivered in multiple media (including Web pages, email, telephone, and print), and players discover story elements, many of which are hidden. Such games often require a considerable degree of collaboration among players. For example, in the alternate-reality game *I Love Bees* (2004), a viral marketing campaign supporting the release of video game *Halo 2* (2004), a series of story pieces were released one at a time through various media all over the United States and parts of the rest of the world. Some of the story clues were released in different languages, requiring some collaboration among players to decipher. A key feature of ARGs is a "This is not a game" mentality, where the illusion that the game may be real is maintained by the game's designers and often the players, who may not see themselves as players at all. The game is never announced or acknowledged, and players must find their way into the game through rabbit holes that are often not obvious.

In one alternate-reality game, Trent Reznor's industrial music band Nine Inch Nails (in association with the puppetmaster of the game, the immersive entertainment marketing company 42 Entertainment) created an ARG to promote the release of a new concept album, *Year Zero* (2007). *Year Zero* presents a dystopian vision of the future United States and is a thinly veiled criticism of the administration of George W. Bush. The MP3 format lacked the album packaging that historically accompanied concept albums, which meant that Reznor's concept album required a rethinking of the format: "So I started thinking about how to make the world's most elaborate album cover . . . using the media of today" (quoted in Rose 2007).

In February 2007, a Nine Inch Nails fan discovered a hidden message in a Nine Inch Nails tour shirt—the phrase "I Am Trying to Believe." Using the phrase as a Web address or uniform resource locator (URL), a Web site was discovered that described a strange "presence," a hand reaching down from the skies. The Web site questioned whether the vision was a form of mass psychosis that was induced by the drug Parepin, which was introduced into the American water supply, allegedly to strengthen the human immune system against biological attack. Using the URL's Internet protocol (IP) address as a base, fans discovered a number of related Web sites—Another Version of the Truth, Be the Hammer, 105th Airborne Crusaders, and Church of Piano. Each of the Web sites described a paranoid dystopian near-future world. Sending an email to the contact address on the I Am Trying to Believe Web site resulted in an email auto-reply: "Thank you for your interest. It is now clear to me that Parepin is a completely safe and effective agent developed to protect us from bio-terrorism. The Administration is acting purely in the best interests of its citizens; to suggest otherwise was irresponsible and I deeply regret it. I'm drinking the water. So should you."[20]

Two days after the t-shirt discovery, a USB flash drive containing the leaked track "My Violent Heart" from the forthcoming album was left in a bathroom stall at a Nine Inch Nails concert in Lisbon. The USB turned out to be the first of many such planned leaks, which were a means of spreading the game at each concert (some USBs were never found, according to Bonds 2011). A t-shirt from the Lisbon show revealed highlighted tour dates, and the numbers revealed a Los Angeles phone number. When called, a clip of a Nine Inch Nails song ("Presidential Address: America Is Born Again") was played after the message. A fan scanned the static at the end of the "My Violent Heart" MP3 file in a spectrogram program and discovered an image of "the Presence" described on the Web site (figure 3.4). A second USB was found a few days later in Barcelona: it contained

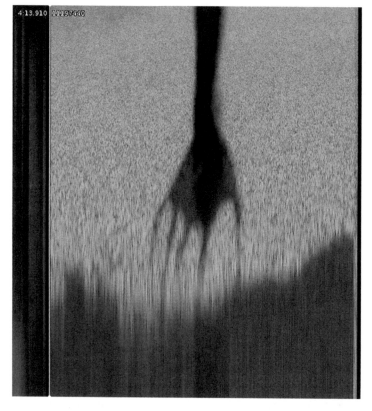

Figure 3.4
Spectrogram of the Presence hidden in the Nine Inch Nails "My Violent Heart" MP3.
Source: Image from *Year Zero Experience,* http://yearzeroexperience.blogspot.com/2007_02_01_archive.html.

another track and an MP3 file of blips of static. Once again, the static was run through a spectrogram, and a phone number was revealed. When the number was called, the player heard screaming and gunshots, and it led to yet another Web site containing the transcript of a wiretapped conversation.

A few days later, a trailer was released on the Web site for the *Year Zero* album: watching the video frame by frame led to the discovery that a single frame revealed a secret link to the album cover art. The same day, fliers were handed out at a concert in Paris with the words "Art Is Resistance." Visiting the "Art Is Resistance" Web site led to a screeching static of noise and another harsh warning, but if the player put up with the screeching noise, links were revealed to stencils and other artwork that fans could download to "spread the resistance."

Hidden messages were also found in the album itself when it was made available: Morse code, for instance, pointed to another Web site. The CD was heat sensitive, and when the disk was played and therefore warmed, it revealed binary code that led to yet another Web site with further clues. A warning on the album read, "consuming or spreading this material may be deemed subversive by the United States Bureau of Morality. If you or someone you know has engaged in subversive acts or thoughts, call: 1-866-445-6580." Calling that number led to another ominous warning:

This is a message from the United States Bureau of Morality, pursuant to statute 24.12.2, Disclosure of Surveillance. Citizen, by calling this number you and your family are implicitly pleading guilty to the consumption of anti-American media and have been flagged as potential militants. The United States Bureau of Morality has activated the tracking system embedded in your personal media and initiated citizen surveillance. United States surveillance law gives us the right to search and seize information relating to subversive activities from your person, vehicle, workplace, or home. Any attempt to hinder or prevent our investigation will be met with all necessary force. You are now part of the problem. Your reeducation is about to begin. God bless America.

Other Web sites were found as fans searched through leaked videos, fliers, and other material put out by the band, gradually revealing a story of a corrupt autocratic government that was drugging its citizens and a citizenship that was affected by global warming and a number of foreign wars.

The game ended in April 2007, when the game became more specific to a single location, Los Angeles. Fans were invited to a special "Art Is Resistance" meeting where they were given a special kit that included prepaid cell phones. A few days later, fans were called on the phones and invited to a location for a free concert. The entertainment marketing company 42 Entertainment remarked on the trust of the fans, who were required to give up their phones, not tell anyone where they were going, and board a bus to an unknown location (Bonds 2011).

As with most dystopias, the world depicted in *Year Zero* resembles the present day. Themes common in dystopia are repeated here—the drugged populace (e.g., Huxley's Soma), the omniscient overlords (e.g., *The Matrix*), mass surveillance (e.g., *1984*), the control of mass media and other communication channels, propaganda (e.g., *V for Vendetta*), and a critique of religion (e.g., *They Live*).[21] Fans of the band were probably familiar with dystopia's tropes and with video game play, if not alternate-reality gaming. Reznor had scored the music for game *Quake* (1996). Thus, Reznor used the game as a way to situate players in the position of the stereotypical dystopian hero, the outsiders who overcome the oppressive government.

One important factor in the success of the game was that players were involved in the coproduction of content and meaning. Fans downloaded and shared the provided Art Is Resistance posters, took the message to heart, and produced their own resistance art. By creating and sharing the artwork, fans became the Resistance movement of the narrative. They became not just players of the game but characters in the game through their active role in producing content that became part of the gameworld. Fans were not just an army of game characters fighting a corrupt government: they became a propaganda machine for the band in a symbolic, simulated resistance that simultaneously buys into the system that it allegedly opposes.

Other alternate-reality games have encouraged cocreative and collective play as the means through which the story is told. *I Love Bees* was presented as a kind of radio drama and was gradually released to its audience through 40,000 pay-phone calls, Web sites, blogs, emails, MP3s, and more, providing a back story to *Halo 2*. In this way, "The distributed fiction of *I Love Bees* was designed as a kind of investigative playground, in which players could collect, assemble, and interpret thousands of different story pieces related to the *Halo* universe. By reconstructing and making sense of the fragmented fiction, the fans would collaboratively author a narrative bridge between the first *Halo* video game and its sequel" (Sean Stewart in McGonigal 2008, 202). As the project's lead writer Sean Stewart explains: "Instead of telling a story, we would present the evidence of that story, and let the players tell it to themselves. . . . The game isn't the art, or the puzzles, or the story. They are designed to precipitate, to catalyze the actual work of art. Which is you" (in McGonigal 2008, 202). The *Year Zero* game allowed the players to experience music in a new way and to become a part of the collective legend of the music and the band. Through such collective play, another story is created that supplants the master narrative, truly allowing the players into to the story (McGonigal 2008). The ambiguities that are often inherent in such games allow players to create and establish their own meanings. Players learn from each other and reduce ambiguity through the collectivity of this meaning generation. Through collective play, the meaning of the music is not just revealed to fans but also created by fans. By allowing fans to enact a role inside the narrative of the music, fans are creating that narrative. The role-play places fans in the role of both consumer and actor in the story, drawing them into the music and letting them experience the music through that enactment. In this way, fans become part performer and part fan. This productivity helps to reduce the perceived distance between fans and a band as successful as Nine Inch Nails, creating a sense of intimacy with the artists that in other forms of music consumption is unlikely.

Fans also become more intimate with the music when they are involved in cocreative practices. In this alternate-reality game, fans were encouraged to play the music and also to play *with* the music by running it through spectrographs and filters, listening to it forward and backward, obsessing over potential keywords and clues in the lyrics, and so on. The game thus facilitated a dialog about the music between fans that extended well beyond whether fans liked songs and into discussions of lyrical meaning and musical structure. By enticing fans to listen actively and engage with the music, the game encouraged new ways of interacting with that music.

New Sonic Boundaries: Identification, Performance, and Cocreativity

In this chapter, I have described three different means by which role-playing in games allows players to experience sound and music in new ways and by which sound enables players to experience games in new ways. Sound acts as a mediator between fan and creator, player and character, individual and collective, reality and fantasy. Interactive sound— whether voice, sound effects, or music—can allow the player into the game-space, encouraging them to take on the role of their character. Interactive sound thus facilitates and enables role-play in games, whether as the subject of the physical enactment of the game's character in alternate-reality games, the performative agent of music-based games, or the voice of the character in online games. Through interaction with sound, players thus adopt and act out the identity of the character.

Interacting with sound encourages bodily engagement with games. Even where players might (to some extent) transcend their physical bodies in virtual worlds, through the use of the voice they can reengage that body and bring elements of reality into the virtual space. In other words, the voice and the sampled sound effects that are evoked bring the auditory fidelity of reality into the virtuality of the game. Because digital games are entirely constructed spaces, this sonic realism can be the means by which a sense of believability is formed. Through the synchresis of the sound effects with virtual visualized objects and the synchresis of the player's voice to an avatar, these virtual objects and beings take on new emergent meanings and gain believability.

The types of interactive sound discussed in this chapter also encourage cocreativity with games. Through in-game chat systems, players engage in meaning creation and meaning sharing. Players create game content through such chat. Whether adding a layer of auditory energy to a virtual space, engaging with other characters through chat, or adding to or altering narratives, voice can become an important factor in generating unique experiences for the players. In alternate-reality gaming, information sharing and cocreativity become vital parts of the experience of the game, as the game adjusts and unfurls in real time, depending on the activity of its players. And through performative activities, interactive sound enables role-play and cocreativity, encouraging social interactions that span beyond the game itself. But cocreativity involving sound extends well beyond the types of play discussed in this chapter. The many other types of cocreative practice with sound in games are the subject of the final two chapters.

4   Embodying Game Sound in Performance: The Real and the Virtual

Video games are a performative and social activity that often extends well beyond the playing of a game. As discussed in chapter 3, games are a highly social medium in which players play against or with each other both online and also in the same physical space.[1] The social interaction that takes place in and around games is one of the most commonly cited reasons for playing games, with one study finding that three quarters of players play with other people (BBC News 2008). Twenty million players have reportedly spent 17 billion hours on Xbox Live, the networked play component of the Xbox360 console. PlayStation has another 40 million PlayStation Network accounts (Pixl Monster 2011). One of the reasons for the growth of social play (in addition to voice over IP) is the introduction of controllers that afford more social interactions and affect social behavior during gameplay (Lindley, Le Couteur, and Bianchi-Berthouze 2008). In this sense, the game becomes a site of performance, which can be defined as "embodied acts, in specific sites, witnessed by others (and/ or the watching self)" (Diamond 1996, 1). Talmadge Wright, Eric Boria, and Paul Breidenbach (2002) describe massively multiplayer online games "platform[s] for showing off human performances" where play extends far beyond playing within the rules of the game. These are "reputation games" in which people dress up their avatars and show them off in public spaces in the game. Indeed, "without an audience of other players to whom these items could be displayed, the game would make little sense" (Ducheneaut et al. 2006, 413).

In addition to the in-person performance and spectacle of gameplay, players also perform in online games and virtual worlds through various antics such as impromptu dance performances. This spectatorship function is built into some games through the design decision to include humorous objects or actions through emotes—text-entry keywords that describe an action that a character can take. *World of Warcraft* (2004), for instance, has emote commands "/dance," "/silly" (in which a random humorous phrase is uttered), and "/violin," which enable the player to play "the world's smallest violin" for a collaborator or opponent who whines. By enabling these emotes, the game developers encourage these performative activities in the game, allowing for functions that span beyond what is necessary to play the game (dancing does not kill any monsters) and into the arena of social interactions.

Several authors have situated the playing of all games as a type of performative activity, with either the player or the character (player proxy) as the actor. According to Emma Westecott (2009, 2), for example, the game character's role is similar to that of a puppet, and the player acts out the game through this actor-puppet: "Closely connected, play becomes performance via the game screen. Digital gaming always involves a screen, producing a doubling in which actions on a controller are represented back on screen. Thus the player is always audience to her own play act. She progresses through a given game always watching the results of her actions on a screen that shows an ever changing theatrical performance built by code and run by numbers." Others, however, suggest that the player's role is closer to that of a director who oversees the character-actors, particularly when it comes to games "in which there are no dramatis personae," like *Tetris* (1984) (Solidoro 2008, 58). As such, "The player performs the role of a character central to the videogame story, since the game itself cannot progress if the players refuse to play the role assigned to them. . . . Spectatorship and active performance are here, surely, merely functional categories and they are not mutually exclusive since most videogames offer the player both the role of passive audience and that of 'actor'; and the player's role may vary not from videogame to videogame, but also from scene to scene in the course of the same game" (Solidoro 2008, 57). Part of the difficulty in delineating the player's role in the performance of games is that the role often changes within a single game: players may move from first to third person and from interactive sequences in which they can control the character to noninteractive sequences where they are only spectators.

Often neglected in studies of the game experience, spectators may be directly or indirectly involved in the game, which can be an important part of the social experience of game play. The term *spectator* here refers to people who may be watching (and listening) as a game is played but who are not physically interacting with the game. Passive spectators might be watching or listening while doing something else, active spectators are actively watching and listening and perhaps creating their own meanings from the activity, and interactive spectators actively help the player in accomplishing goals by drawing maps, keeping track of objectives. James Newman (2008) differentiates between what he terms "on-line" and "off-line" engagement with the game, where *on-line* refers to playing the game (physically) and *off-line* play refers to the times when there is no registered input control but there is still involvement in the game. Thus interactive spectators may be physically off-line while aiding the on-line player. T. L. Taylor and Emma Witkowski (2010) refer to this secondary role as the "backseat player" and argue that spectators are not merely observers or bystanders but are often actively engaged in the game and thus engaged in another, less direct form of play.

In other words, even single-player games can become a social, multiplayer activity through spectatorship, and a wide variety of social activity takes place with, in, and around games. In this way, "Often simply watching a familiar game connects you, somehow viscerally, to your own embodied experience of play. . . . You may second guess an action, remember your own prior experience of playing that scenario, be awed by some new action you are seeing, or be moved to go back and progress further or relive the gameworld. . . . It can reground your identity as a gamer and even viscerally pull you into that play moment, sometimes even transforming it into a kind of shared experience" (Taylor and Witkowski 2010).

The intersection of play and performance in games is complicated and raises interesting questions regarding digital music performance and liveness, as well as the role of embodiment in these performative activities. As was shown in the previous chapter, game sound itself can become a form of performance in which players act out the character in elaborate role-play through the emulation of musical gesture and articulation in rhythm-action games and through virtual performances that take place in the game space. In this chapter, I explore how sound mediates other types of interactions between game players and spectators, as the player engages in a variety of performative roles. Game sound, in this sense, becomes both a tool for and a site of performance and spectatorship. I first investigate the ways in which players perform music *in* games, exploring notions of liveness and embodiment in virtual performances. From there, I examine ways that players perform to music in games, using the sound as an opportunity to engage both the physical and virtual body. Finally, I explore the performance of music outside the context of gaming altogether—the use of game sound and music as components of musical performance—and examine how players become involved in performance and textual productivity through the recontextualization of game sound.

Performing Music in Games

There are a variety of mechanisms through which players may perform music inside a game or virtual world, including streaming music into the space and composing music in-game. As discussed in the previous chapter, although the original idea of voice over IP was to communicate and share tactical information during gameplay, many players use voice chat to socialize, and some players stream music (over the headset microphone) into the game, sharing songs with other players. This practice became so popular that Teamspeak designed the idea into the third version of its software to make it easier for players to share music over the network through an artificially intelligent agent known as a "bot." The desire to set up a personal streaming channel has led to a kind of in-game radio system where players can broadcast their own music, effectively becoming in-game DJs. However, this streaming of music into the game sometimes dismays others. Many people like to play their own music selections, but few like to be forced to listen to someone else's music. Competing VoIP software Ventrilo, for instance, warns users "it can be annoying for people who don't want to hear it, especially if your server is for a guild that raids or works together in a massive multi-player online game. For these purposes, one option is to create a channel for music that others can access and play in" (Darrington 2011).

Some games have allowed player-characters to play a musical instrument in the game, although they may have little control over that music being played back. For example, in *Asheron's Call 2: Fallen Kings* (2002), different species of characters play different types of instruments, and the interaction between players lead to changes to the music (see Fay, Selfon, and Fay 2004, 473–499). Recent systems have allowed for more advanced musical creation ability, allowing players to create and perform their own music in the game. Casual MMO game *Glitch* (2011), for instance, allows players to collect musical sequence blocks and then use whatever blocks the player has collected to construct a song. In this way, the overall music is produced according to the player's individual experience of the game, even though the sequences themselves are pregenerated.

In some cases, players can compose music by pressing music-mapped keys or inputting an ABC Notation file to create and perform their own music in virtual worlds, such as in *The Lord of the Rings Online: Shadows of Angmar* (*LoTRO*) (2007) or *Mabinogi* (2004).[2] *Mabinogi* allows players to compose their own music in the game using music markup language (MML) code to write score scrolls that can be used in the game. The ability to compose music was integrated into the game in that players can assign their character points to become better skilled at composing and collect coins, buskering in the space. Nevertheless, allowing players to compose their own music has inevitably lead to copyright questions (since players have reproduced popular songs in the game), and there are rumors that the game's creators are removing the composition system in the wake of threatened changes to copyright law.[3]

LoTRO allows players, once they have obtained a musical instrument, to enter a "music mode" that uses the ASCII keyboard to play songs in real time. Although many players like to explore the musical option, it can be annoying to other players in the game: "That is the one system about *LoTRO* I dearly hate. Nothing like standing in 21st hall and having some moron pull out a set of bagpipes and start playing 'Freebird' or something else. That is enough to make your ears bleed" (ericlewis in Webb 2010). Some players feel that the music function brings nothing of substance to the game: "I thought it was a very interesting innovation. I had some fun with it, but of course, those with no musical ability are annoying to be around. It would have been interesting if playing certain notes had some use in some quests, but as a novelty item, it gets old after a while. It's probably not copied because for the effect required to implement it well, there is limited benefit" (dadown in Webb 2010). On the other hand, many players like the system's innovation and enjoy the ambiance and excitement that it gives to the game, particularly the feeling of a lively social atmosphere: "I love the music system in *LoTRO*. I enjoy hearing people standing around playing instruments. It makes the towns more interesting and alive. *LoTRO* is the only game I've played where people sometimes sit around in a tavern, talk, play music, and smoke pipes for fun" (Elirion in Webb 2010). The important idea here is that even with mistakes (or perhaps because of the mistakes), player-generated music "makes the towns more interesting and alive." Contextually awkward music (such as "Freebird") may disrupt the immersion for players who are familiar with the songs but also allows an outlet for players who want to express themselves musically. Player-generated music, in other words, may recognize its own artifice and may break with the immersive quality of the virtual space, but potentially it creates a different type of engaging experience for players.

With some in-game musical performances, such as *Second Life* (2003), the music may be entirely player-generated and streamed into the virtual world space. Performances are created by streaming music live from the performer's computer through Shoutcast or Icecast[4] and then tied to a

Figure 4.1
Live performance in *Second Life* (2003).
Source: Image from http://www.flickr.com/photos/raftwetjewell/5577615757/in/ photostream.

section of land or to one of the clubs in *Second Life* (figure 4.1). Landowners can also stream music to their land, essentially creating a soundtrack for their own space. At some of the *Second Life* concerts, the audience attempts to tie the lyrics to avatar actions. Craig Lyons, for instance, performed a song called "Winter," and the audience made it snow, and Maxx Sabretooth's song "You Can Leave Your Hat On" sent fans searching their inventory for hats to wear (Ferreiro 2010). David Cameron and John Carroll (2009, 3) argue that such live performances "strengthen the sense that this is a more traditional performance, albeit totally mediated through virtual environments and characters. Unlike a real-world stage performance, the software is creating the lighting, sets, and characters on the fly in real-time; but the action and plot and performances unfold at a human pace." Nevertheless, these types of virtual performances call into question our notion of what it means to perform as a musician and what it means to perform live. If players use an avatar as their visual representation, are they really performing for the audience? And if that audience is only virtually present, is it really an audience? When players are singing live in a bedroom but their avatar is performing prescribed moves to that music in the virtual world, is the players' performance really live?

Online games and virtual worlds are now often being used to comment on the nature of avatars as players' disembodied selves. Reperformances of "real" live events in the virtual space illustrate the interesting ability of virtual performances to comment on the notion of the body and liveness. Eva and Franco Mattes, for instance, re-created famous body art performances of the 1960s and 1970s, including works by Joseph Beuys, Chris Burden, Vito Acconci, and Marina Abramović in a series called *Synthetic Performances*. By re-creating performances that originally focused on the physical body as the medium for the message, we are forced to reevaluate the notion of bodies in the virtual space and question the ways that we extend our sense of self and body schema into that virtual world. If, as research suggests, real-world physical boundaries like cultural proxemics carry forward into the virtual (Blascovich and Bailenson 2011) and if, as I suggest in chapter 2, we can extend our sense of self into the virtual space, what does it mean to have a body in this space?

According to the embodied cognition theory of mirror neurons and the concept of kinesthetic sympathy described in previous chapters, the ability to understand a performer's emotions lies in our kinesthetic sympathy (mental mimicry) of performing such actions ourselves. Currently, avatars are somewhat limited in their physical emotional expressiveness (particularly when it comes to subtle gestural articulations) and are even further limited in their ability to move each of their limbs as a performer may want, beyond merely typing an emote command. However, the visualization of an emotional performance that uses avatars does not necessarily disrupt players' ability to mimic the emotion that they hear. They are hearing the real performance, after all, even if they do not see that performance. Although I have found no research to support this contention, I suspect that we mentally fill in the visual gaps that exist in the avatar's representation of the real sonic performance. An ability to understand the performance, in other words, lies in our listening ability rather than our sight. Performance in the virtual space may strip the visual embodiedness of performance, but that embodiedness continues to exist in the sound. Paul Sanden (2009, 9) notes, in relation to synthesized sound, that "Arguably, the important thing is not whether sound was actually physically *produced* by a living being, but whether we *perceive* some sort of live presence in those sounds." In a similar manner, what matters in terms of empathic experience is not the visualized form of the avatar but the perception of a real performer behind that avatar, a realness that is granted by sound.

Melanie Fritsch and Stefan Strötgen (2012) distinguish between a "live performance" and a "live music performance"—the latter referring to the live sound and the former referring to a live musician on-stage. Just as a live musician can perform without live music (such as by lip-synching), so can live music be performed without a live musician on-stage. Virtual bands like the Gorillaz, for instance, can perform live without being physically apparent to the audience. Virtual performances in games can likewise be live musical performances, even if the musician who is seen is not the physical body of the musician.

Philip Auslander (2002) suggests that our concept of liveness is fluid and articulated in relation to technological change. It was not, for instance, until the advent of recording technology that the notion of liveness was brought into being. with the arrival of broadcast technology (radio), this relationship was called into question. He suggests that "As a consequence of the circumstances under which this vocabulary was instated, the distinction between the live and the recorded was reconceived as one of binary opposition rather than complementarity" (Auslander 2002, 17). Auslander (2002, 20) makes an interesting case for liveness in his discussion of the performance of chatter bots, artificially intelligent conversationalists that perform in real time. In the same way, *real time* has become synonymous with *liveness* in virtual worlds, and the notion of liveness is being adjusted with this new technology. If it is accepted that a performer's sound is live (and streamed into the virtual world) and that the avatar's performance (being in real time) is live, then there can be a simultaneous dual performance—one the physical embodied live performance of sound and the other a virtual visualized live performance of the body.

Performing to Music in Games

In addition to video games and virtual worlds that allow users to incorporate streamed music and thus to participate in in-game virtual performances, games have been used to create performances that are then set to music. These are often recorded in machinima videos. The word *machinima* was coined from *machine* and *cinema*. Machinima movies are made from video game engines and include many different types of practice, including puppeteering for an audience (performing through the avatar), demos and speed runs (game engine replays of recorded playing performances), recamming (altering the game engine replay), scripted bots (avatars that are programmed to behave in a certain manner), and video files captured and edited in postproduction. Machinima evolved in part from the history of *vidding*, in which television and movies are mashed up and set to music, with the result known as a *vid* or *songvid*. The history of vidding can be traced back to a 1975 *Star Trek* slideshow that was set to music and called "What Do You Do with a Drunken Vulcan?" (Coppa 2008). A second influence is likely that of MAD movies, which are anime parody videos that are made primarily in Japan and were popular in the 1980s (Ito 2011). These videos were shown at fan conventions and had limited distribution until the arrival of YouTube (Jenkins 2006b). Particularly relevant to this discussion and to machinima's development is the important role that has been played by music in the construction of vids and MAD movies: many were montages set to music, in which "vidders use music in order to comment on or analyze a set of pre-existing visuals, to stage a reading, or occasionally to use the footage to tell new stories. In vidding, the fans are fans of the visual source, and music is used as an interpretive lens to help the viewer to see the source text differently" (Coppa 2008). MAD videos, for instance, often swapped out the original song with a new song whose rhythm and lyrics coincidentally aligned with the original. Songvids, as music-based vids became known, are arguably the most popular form of vidding and influenced machinima.

Machinima developed in parallel to the demoscene and modding, both of which hacked elements of the game engine to allow for some degree of user control (see chapter 5). By the early 1990s, some game engines like *Doom* (1993) enabled users to record gameplay as data that the engine could later replay. Users could share these data files with each other, and so demonstrations of gameplay techniques were shared. These game replays helped players to learn skills and strategy and to show off their prowess with a game. Although interactivity is removed from the demos, the experience of watching them is not quite the same as video. As described by Cameron and Carroll (2009, 3),

If the player moves their character forward and shoots a weapon, the character's changing co-ordinates in the 3D space and the player's command actions are recorded and stored as data in the demo file. . . . When the demo file is replayed, the software engine can replicate the same input over and over, feeding the data to the game world and characters. . . . Yet there is a contextual and experiential sense in which the viewer is aware that this is not an animated film. . . . As technically precise as many of these virtuoso demos are, there is still a sense that you are watching a human generated performance.

With the release of *Doom*'s sequel, *Quake* (1996), recamming (altering a demo movie with different camera views) and postproduction editing were added to the original concept. But because the files that were released as data code required a copy of the original game to play back, *Quake* movies, as they became known, were limited in their audience. Moreover, as networked play arose in the mid-1990s, moviemakers shared information on how to cheat the games, and the makers of *Quake* (id software) locked down the code for *Quake III Arena* (1999), leading to the demise of the *Quake* demo format around 2000. With this demise, however, came the birth of mods, such as *Unreal* (1998) and *Half-Life* (1998), which were based on other game engines and allowed the files to be distributed as either data logs or as encoded movies (Lowood 2008). A distinct advantage of machinima as video file rather than code was that its makers did not need to be programmers but could use the game engine to render and record movies using video capture. These video files also could be more easily shared with those who did not have copies of the game.

As with more conventional movies, machinima videos today come in many different genres and often are dependent on music. There are some abstract machinima that are marginal to the community due to lack of perceived narrative (Cannon 2007, 43). As with songvids, many abstract machinima are set to music, such as *gLanzol* and *Fabelmod*, which are modifications of *Half-Life* (1998) and *Counter-Strike* (1999) and use modified game engine code as a backdrop to electronic music live shows, essentially using the game as visual performance (figure 4.2).[5] Similarly modified "data bending" like *Carmageddon* edited the actual data of car crashes from

Figure 4.2
Fabelmod by Glaznost: An art mod conversion of the *Half-Life* (1998) game engine.
Source: Screen capture from video http://vimeo.com/24964394.

the *Carmageddon* (1997) game to morph into a series of polygons set to a noise soundtrack. The creator, Cementimental, is active in machinima, circuit bending, and experimental noise music scenes, illustrating the cross-over among the communities. According to Cementimental, "These movies were created by editing the data which determines the crumple behaviour when the cars in the game crash. By inserting ludicrously large figures, and then repeatedly triggering the 'bodywork trashed' powerup via a cheat code, the player car is mutated into a jagged mass of mangled polygons which fill almost the whole screen, and becomes a moving virtual abstract sculpture."[6]

These more abstract machinima use game engines and game scenes to provide a visual accompaniment to electronic music. One of the difficulties with performing electronic music live is the (perceived) lack of visual interest because the emphasis is removed from the body of the performer and placed onto the music itself. With electronic music, the sound is not limited by the skills or physical limitations of the performers, and thus "need no longer bear any relationship to anything that can be performed live" (Simon Frith in Théberge 1997, 76). Performing live often poses problems for the electronic-based artists due to the amount of studio-based processing, the speed at which some instrument sounds must be played, and especially the limitations of keyboards in allowing stage movement. Without an engaging physical performance to watch and the visualization of sounding bodies behind the performance, audiences often feel that electronic music requires some accompanying visuals to combat the "audiovisual disjunction"—the schizophonia (described in chapter 1) in which listeners attempt to fill in the visual gap ("the menacing void") in music recordings through the production of visual accompanying material of music videos, album covers, and so on (Corbett 1990, 84). In other words, machinima videos used as visual accompaniment to live performance can become a form of synchretized performance in which a new fusion of audio and visuals can be achieved in the absence of much physical movement by the performer.

Another musical genre of machinima is the creation of music videos, which include live performances of artists' own music as well as constructed creations that are set to popular songs. Virtual music performances that have been prerecorded and rendered in video format are fairly common. Blizzard Entertainment, the creators of *World of Warcraft* (2004), produced a machinima video of its employees as characters from the game performing "I Am Murloc," a song written by Blizzard employees performing as the band Level 70 Elite Tauren Chieftain, among other names (figure

Figure 4.3
The Elite Tauren Chieftains performing in Blackrock Depths.
Source: Level 90 Elite Tauren Chieftain *WowWiki.* Available at http://www.wowwiki
.com/The_Artists_Formerly_Known_as_Level_80_Elite_Tauren_Chieftain.

4.3). The song became so popular that it was later released as a *Guitar Hero III* (2007) track. Although this particular Murloc video was not recorded in-game, to some extent it carries the illusion of being made in-game. Edda Stern (2011, 43) says that

Creating "Murloc" as machinima rather than as a slick rendered animation was a sophisticated marketing device: a type of corporate "fakelore," an artificial cultural artifact produced in such a way as to resemble a real piece of folklore. In this case the rough machinima technique produces a seemingly authentic product. Blizzard has deployed the vernacular language of machinima and given a savvy nod to its community by sharing in the production of a "low brow" yet hip media artifact, effectively fortifying the company's credibility within its consumer/fan base.

In addition to recordings of virtual live performances, it has also become popular for machinima creators to use popular songs and enact some element of the song or otherwise create a video for it. The most popular form of machinima music video is to create new videos for these songs—such as Toto's "Africa" using *The Sims* (2000), Weezer's "Beverly Hills" using *Halo 2* (2004), and David Bowie's "Space Oddity" using *The Sims 2* (2004)—although cover versions and remakes that emulate existing popular music videos are also common. Brian Turner's "Synthetic," for instance, borrows from the original Chris Cunningham video for Icelandic singer Björk's song "All Is Full of Love." The original video used two android Björks in various stages of homoerotic affection and has often been discussed as a commentary on virtuality. For Tanja Shaviro (2010, 22), "Björk and Cunningham do not critique virtualization, so much as they open up its potentials. They (re-)find or rediscover the body at the very heart of virtual reality and cyborg-being." According to Shaviro (2000, 27), the original video's appeal is due to Björk's own disembodied, digital voice, which questions the nature of what it is to be human:

It is a double movement, a double seduction. On the one hand, Björk's voice is dehumanized. It sheds the richness of texture and timbre that individuates a singing voice. Instead, it tends toward the anonymity and neutrality of digital, synthesized sound. It becomes less analogue, less vital, and less embodied. The living person moves closer to being a machine. But on the other hand, and at the same time, the nature of the machine is also transformed. At the heart of this digital blankness, a new sort of life emerges. Precisely because Björk's voice has lost its humanistic depth, it is now able to float free. Spare and without qualities, it is open to the minutest fluctuations of rhythm and tone. The voice wavers and hovers, on the very edge of perception. In this way, it weaves itself a new, tenuous body. At the same time that Björk herself is recast as a digitally programmed android, the digital machine itself becomes more analogue, and more nearly alive.

Brian Turner's machinima take on the video uses differently gendered avatars from *The Sims 2*, in an odd parody that manages to comment on both the Björk video as well as players' lives in the virtual world. Here, the robots are supplanted by an avatar human and an avatar robot as a virtual male manufactures an ideal mate—a virtual android female that is reminiscent of the way that players can create avatars onscreen. The love scene that takes place here is therefore not between two androids but between fake human (avatar) and double fake human (avatar-android) (figure 4.4). In this way, Turner's video becomes a metacritique of the meaning of virtuality and humanity. Viewers identify with the human avatar in the video, and yet it is as virtual as the robot, so what distinguishes viewers from these virtual creatures?

Other fan vids have created their own songs or at least lyrical additions to popular song and have used them as the basis for machinima music videos. Part of *Goodbye Christmas Caroline*, a Christmas album of songs featuring characters from Valve games, the *Portal 2* game encourages fan remakes because at the end of the game, a chorus made of robot Turrets from the game sing the "Turret Opera." A Christmas remix of the song

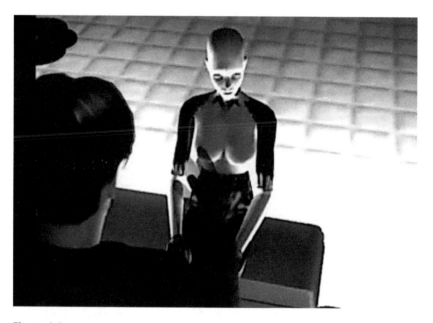

Figure 4.4
Screen capture of the video for Brian Turner's "Synthetic," a take on Björk's "All Is Full of Love."

"Still Alive" from the original *Portal* game was also released by Valve. Although not actively encouraged by Valve because the intellectual property remains copyrighted, fan remakes of Valve songs have not been pursued in court, as the creator of the "Turret Opera," Jonathan Coulton (2011), describes:

It is not under Creative Commons. Technically you would have to get permission from Valve to do something with it. In the past they've not seemed to mind all the crazy things people did with *Still Alive*, so if I had to guess (and I do), I would say that as long as you weren't profiting from it in some way, they'd generally be supportive of fan-created stuff. But this is not legal advice, and I certainly can't give you (or refuse you) permission to do anything.

"This Is Aperture," a parody of *Portal 2* with a musical remake of Danny Elfman's "This Is Halloween" (from *A Nightmare before Christmas*), is a musical machinima that uses the characters and scenes from the game.[7] Creator Harry Callaghan did all the robotic "GLaDOS" (Genetic Lifeform and Disk Operating System) voices of the game using Auto-Tune and distortion effects. Providing a humorous warning to the player, the song presents the basic storyline and introduces the key character GLaDOS:

GLaDOS the giant may fill your room with neurotoxin!
Shoot you with a rocket-turret—turn you into burning mush!
This is Aperture!
Everybody RUN!
You better escape to a much safer place!
Our dear Chell is master of the portal-gun!
Everybody help her in her flight!
NOW!
This is Aperture![8]

The use of virtual world performances and recordings to comment on embodiment, performance, and liveness suggests that creators are attempting to understand how players relate to their avatar body and what it means to have a virtual presence. The curiosity about the sometimes contradictory nature of virtual bodies can be seen in the dance machinima that have also been produced in virtual worlds. One popular *World of Warcraft* video, "Shut Up and Dance," opens with Carl Orff's *Carmina Burana* (1935–1936) as it describes the setting of the scene, showing clips of the landscape and some of the game's characters. As the music becomes particularly dramatic, sound effects are brought in, and fighting ensues. The movie then asks, "What if for one day everybody put down their weapons, and ceased their spells. . . . What if for one day, everyone dropped their armor to the ground. What if, for one day, every soul Azerthoth had no choice but to shut up and dance?" The video then cuts to various characters dancing to the Junkie XL remix of Elvis Presley's "A Little Less Conversation." Various characters demonstrate their dance moves, including gyrating hips, Cossack dancing, disco dancing, pogo dancing, and air guitar (figure 4.5). Such dance moves are built into the game with the emote command </dance>, each race having its own moves and style of dancing. In this way, the creator of the game, Blizzard, allowed for and even encouraged such activities to take place. The "Shut Up and Dance" videos became an Internet meme as other players created their own "Shut Up and Dance" videos in *Sims 2*, *Armed Assault*, and other games. Such videos sometimes require significant coordination among groups of players to perform and record the performance, synchronize moves, and amalgamate the clips together to the song in postproduction.

Beth Coleman (2011) argues that "The rendered movement in 'Shut Up' is modeled on various actual dances, for example, John Travolta's disco diva in *Saturday Night Fever* or MC Hammer's famous 1980s hip hop dance moves. The exchange between real-world histories and those of game-world is part of what makes for the humor of the genre." But even more

Figure 4.5
A screen capture from the video for "Shut Up and Dance," with a character showing off her moves.

than introducing humor, the conventional, familiar dance moves made by the disco diva and pogo help viewers to recreate that experience for themselves. Because they are better able to understand what they witness in terms of their own bodily engagement (discussed in chapter 2) by relying on bodily moves that they have likely enacted themselves, this familiar physicality can transcend virtual space.

With the dance videos, players are not able to control the actual moves of their character. As discussed above, beyond the ability to type an emote command ("/dance"), players cannot control how their character dances. As with cinematic sequences, the physical body of the player must temporarily be removed from the game. Interactivity is lost, and the performance becomes not an expression of the player's embodied connection to the music but a simple prescribed sequence of moves that do not require the player's presence. The inability of current systems to incorporate kinesonically congruent moves into game spaces highlights the divisions between avatar and self, reducing a sense of extended body schema. Thus, at times when an avatar is perhaps at its most physically active, players lose their own physical connection to that avatar. Nevertheless, in the future, such virtual performances probably will be actual representations of how players hear the music to which their avatar is dancing and will be tied to the physical body of the player.

As with other virtual performances, machinima calls into question notions of liveness. Some of these types of videos are performed in real time in the game space, but others are produced nearly entirely in post-production. Real-time productions carry the energy of liveness and are key to some forms of machinima, particularly those in massively multiplayer games. Coleman (2011) argues, "If the dancing in 'Shut Up' were not actually performed (real-time), as opposed to compiled and retroactively produced as most digital animation is, it would have no reason to exist. Dancing presents an extra-value in the communication economy, one that simultaneously describes the human 'inhabiting' of game space even as the robotics of the program, its real-time factor, are exploited. That this is simulated movement—action that is the result of human-computer interface of commands—does not make it any less movement." But the real-time (live) performance of dance in machinima videos is much more complex, particularly since "The only thing that actually moves are lines of code" (Coleman 2011). In some very similar machinima, the productions are carefully staged, and Cameron and Carroll (2009) argue that even though machinima may be recorded and edited in postproduction, the films usually begin with some live gameplay and incorporate virtual elements (such as bots, agents, and nonplaying characters) that also can carry on conversations and actions. Moreover, the real-time rendering of content in the game engine is inherently live: the engine creates lighting, sets, and characters in real time (Cameron and Carroll 2009). Thus the idea of liveness can be fluid in discussions of machinima, where a human action through a proxy agent (rather than the human action directly) is recorded and such human action intermingles with nonhuman action.

Performing Game Music

Players frequently participate in two common performative practices around music in games—covering songs and remixing game sound. A cover version is a new recording of a song "for which musicians and listeners have a particular set of ideas about authenticity, authorship, and the ontological status of both original and cover versions" (Solis 2010, 298). According to Gabriel Solis, not every rerecording of a song can be classified as a cover version: the song must have been popular enough to "exist in the memories of musicians and audiences because of a strong, previous recorded version, and for which authority and authenticity are understood to be shared by the original performer and the covering performer" (315). This notion of authenticity is key to Solis's conceptualization of cover versions, but fans are not committed to his definition. In many cases, a cover version can be a "history lesson" for listeners who do not know the original version, do not normally listen to that type of music (but may subsequently seek out the original based on their enjoyment of the cover), or may seek out the covering artist based on their enjoyment of the original (Cusic 2005). For example, Johnny Cash suddenly became an icon among some fans of Nine Inch Nails after Cash covered the song "Hurt." Covers therefore can "provide an intertextual commentary on another musical work or style" (Butler 2003, 1) and broaden listening tastes.

There are many cover versions of popular game songs. The *Super Mario Bros.* (1985) theme song, composed by Koji Kondo, has been rerecorded on YouTube video by fans playing drums, piano, eleven-string bass guitar, flute, acoustic guitar, twin solid-state musical Tesla coils, radio-controlled car and bottles, two guitars played simultaneously by the same person, laser machine, violin, teeth, ruler, trombone, hand-farting, burping, phone, ocarina, banjo, clarinet, harmonica, tuba, church organ, steel drums, jazz band, bassoon, trumpets, balalaika, marching band, and others.[9] As with the games themselves, the covers become part of a larger participatory, interactive network when the audience ranks and discusses them. Players try to add to the sonic repertoire with a unique offering, thus interacting with others who record cover versions, if only indirectly.

There are also the larger symphonic, industry-sanctioned cover versions of game songs that take place in concert halls and are performed by well-known orchestras. The Malmö Symphonic Orchestra performed a game music concert in 2006 that was attended by over seventeen thousand people. The Los Angeles Philharmonic orchestra has been involved in a number of performances, including a *Final Fantasy* (1997) concert held at Walt Disney Concert Hall in 2004 and the touring Video Games Live series.

Cover versions of game music decontextualize and then recontextualize game music, presenting it to a new audience or to a familiar audience but with new instrumentation and orchestration and away from the context of the game. In this way, the cover can lead the audience to view the original in a new light. For example, the orchestral performances of game music allow a game music audience to be exposed to the symphony and allow symphony fans to be exposed to game music: "If it wasn't for video games I would never have bothered to go to any kind of orchestra in my life," writes one commenter in a review of a performance. Another proclaimed that the show introduced him to video games: "I am not familiar with any of the video game music but became interested when I saw the advertisements. It was an amazing show."[10] Music that was created for early games was entirely synthesized on a sound chip in real time during game-play, buy cover versions can introduce elements of the body into the music. With limited capability to add any articulatory, gestural expression in early game music, such songs were in essence disembodied code. Gestural performance is not the only way to ingrain emotion in music, of course. Musical emotion consists of many different factors, including harmonies, melodic contour, and so on. But songs that are generated based on code rather than physical action reduce sound-accompanying gestures (described in chapter 3) and the ability of the player to hear the performance behind the music. By covering the songs, the player can reintroduce the articulations—the sound-accompanying gestures—and bring the body back into play.

Performers cover video game songs for many different reasons. For cover band Thwomp, the cover is about attempting to recreate the game song accurately and perform it live: "We're all purists about it, really. . . . We really try to make everything absolutely note-for-note. Absolutely exact. We have really high standards for each other" (Arca 2010). The NESkimos, on the other hand, create variations based on the original by adding lyrics, expanding themes, or performing creative arrangements of songs. When asked what their motivation was for covering game music, the band responded that video games are "something that binds our generation together" (Sklens 2005). Other bands focus on the virtuosity that is inherent in playing some of the music, which may never have been originally intended for human performance. In other words, the motivations to create cover versions include reproducing the original accurately for an audience that knows the original and appreciates the reproduction as well as recontextualizing the music through new instrumentation, lyrics, or arrangements.

Cover versions are similar to remixes, which maintain identifiable features from the original song but significantly change the original, particularly the structure of the song. In other words, a remix alters the "recorded paradigm" —the recognizable version—of the original (Moore and Dockwray 2010). Unlike a cover, a remix often uses the separate tracks or sounds as recorded by the original artist. These separate tracks are usually provided to remixers for the purpose of the remix. With game remixes, remixers usually do not have access to the original tracks and therefore rerecord the song or, more often, remix old chip songs that have been transcribed and emulated on old soundchip emulators. Entire albums of remixes and cover versions are released and sold through OverClocked ReMix (ocremix.org), which describes itself as "dedicated to the appreciation and promotion of video game music as an art form." These Web sites attract large numbers of fans who want to create or just listen to game music remixes. Over-Clocked ReMix is so well known that it has spun off its own "remix," called OverLooked ReMix (olremix.org), which bills itself as "dedicated to ridiculous interpretations of video game music and video game culture." The different ways that these Web sites describe themselves (with one referencing the other and expecting the audience to be aware of the reference) highlights distinct differences in motive. OverClocked wishes to use the remix to elevate game music and to share an appreciation of game music as art, whereas OverLooked takes itself less seriously and presents over-the-top fun remixes.

One popular type of game song combines elements of remixing and cover versions by recording game songs and adding elements such as orchestration or lyrics. The song "Little Mac's Confession" by Swedish heavy metal band Game Over, for instance, covers elements of *Mike Tyson's Punch Out!* (1987) music and then adds lyrics that tell a story about game character Mac apologizing to his trainer:

Doc, I know I've let you down
Cause you counted on me
But David beating Goliath
Just wasn't meant to be
What's my Star
Compared to Dynamite?
A quite pathetic fight
Let's keep it clean, punch out!

Other cover versions use similar lyrical narratives that explain a character's feelings or actions or add to an existing game narrative, filling in elements of the original storyline and continuing the game experience and narrative beyond the actual game. Like the science fiction fan practice of *filking*, songs about a game's narrative and characters that use preexisting music are increasingly common in the game music community. The songwriting process of these types of songs requires games that employ many options for lyrical material:

I wanted to make music about something I enjoyed and knew a lot about. As much fun as *Halo* or even *Starcraft*, is to play, there isn't nearly as much material to work with. I mean, *Halo* songs would consist of, "oooh baby, another head shot, I'm so pro," and *Starcraft* would be, "I know the build order, I can see your invisible units, neener neener neener." With *WoW*, there's so much to work with. Talent trees, raids, pvp, gear, gems, enchants, etc. (Letomi in McCarty 2010)

Up until I started parody songs, *World of Warcraft* was one of the only games I played heavily besides *DDR*. *World of Warcraft* is probably one of the easiest games to apply these kinds of artistic methods to. It's got a lot of content, a forever-going story line, and almost everything in it is easy to elaborate on or make fun of. The fundamental part of parody writing or writing in general is just about finding a topic you are passionate about and letting your mind go to work. Anything in *WoW* is feasible to write about; it's just how you write it and how well it turns out. (Cryssy in McCarty 2010)

Game songs have thus been created to pass on information to other players in a modern take on oral history traditions of folk songs. For example, the song "Rapwing Lair" by Shaunconnery details how players can defeat every single boss in the Blackwing Lair region of *World of Warcraft*. The creator describes, "3 months ago, I started a rhyme project titled 'Rapwing Lair,' which was a rhythmic story of how to do every boss in [Blackwing Lair]. It was really just a joke at first, but it evolved into a multitude of styles, impersonations, humor, and rap narration" (in Wachowski 2007). An example verse Razorgore, for instance, is done in underground hardcore style:

Blackwing Lair, zoned in as we phased through
40-man crew, with 8 classes in the raid group
So we know whats in store, MT controllin the orb
Breaking the eggs to bring him out to phase 2
Lets get it happenin, Hunters kitin dragonkin
Running up and jumping down ramps for the fake route
With the last egg down, Razorgore gonna break through
The MC, MT settin the pace to
DPS him down, Line of sight the fireballs
Stay behind in spite before it bakes you
So Off-Tank, stand at your 90-degree angle
So when he breaths, it doesnt conflagrate you
Slow and steady, Razorgore givin his paid dues
Tier 2 bracers from his grave loot
Then we line up, 40-man raid group
To move to Vaels room, BWL Take 2.[11]

Here music has been adopted as a method to convey information, even very specific information ("stand at a 90-degree angle"). In many respects, it would make more sense to write a "walkthrough" (a guided instruction of the game). By using music, the instructions are presented in a more interesting manner and recall the storytelling history of ballad song. The songs thus tell the story and help to bind players together in a community that has its own oral history and folklore. One interview with several songwriters using *World of Warcraft* (*WoW*) as raw material describes the importance of this sense of community and the narrative elements in the game as means to inspire song lyrics: "*WoW* has introduced me to great people. Not only have I made amazing friends, I have opened doors of opportunity and worked with some very popular people in the community" (Rawrbug in McCarty 2010).

Part of this sense of community is introducing the norms and cultural expectations of that community to newcomers. Because most listeners of the music are already frequent players of the game, the songs serve as an inside joke that helps to reinforce their sense of camaraderie. This seems to be the effect of the song "I'm Just a Noob" by Sharm: "I'm just a noob / I'd rather not be / but they won't let me quest here in peace. I'm just a noob / guess I'm some kind of freak / cause they all laugh and kill me with ease."[12] Such songs delineate a distinction between those "in the know" and others who have yet to learn the conventions of the community.

Many of these game filk songs are set to popular music that is not taken from video game soundtracks. In this way, the song is tied to the gaming world only through its lyrical content and risks bringing its own semiotic baggage from outside the game community. Songs risk taking on an unintended double meaning by being tied to other textual material (such as the original music video or performer), but the use of popular well-known songs also "facilitates more immediate audience participation" (Jenkins 1990, 162). Another difficulty in using popular songs is the sometimes perceived lack of authenticity associated with not writing one's own music. In discussing cover songs, Don Cusic (2005) writes that most critics and fans of music tend to imbue singer-songwriters with a greater sense of legitimacy than those who do not write their own songs. Bands that do not record their own music are often dismissed as inauthentic. However, Cusic (2005, 172) explains that this attitude ignores the ability of bands to interpret a song in a new way and thus create a different form of authenticity. By changing lyrics, instrumentation, orchestration, and other pieces of the original recorded paradigm, these filk songs can make a claim of originality.

Filk songs can extend the players' identification with characters beyond the game. Such songs are often sung in first person as the character and rarely as the player (for example, the character and not the player is to "stand at a 90-degree angle"). It is another form of role-play where players might take on the character's persona and act out their experiences. Moreover, unlike the character identification that occurs with television and movie filk music, game filkers can voice their own character from the game. Movie and television filk music may speak through a character's mind, but the singer must select a persona to adopt, and the character is never their own creation. With game filk, creators do not need to choose a persona of a character to adopt because they have already created or adopted their character, and in the case of VoIP-supported games, they have already voiced that character with their own voice. Filk therefore is another means by which players can extend their self into the game world through sound by extending their character beyond the game world and by role-playing that character in real life. The music thus furthers their ability to identify with and role-play their game character and interact with other players through sound.

Creating Music from the Game

Besides using games as lyrical content for music, the software and hardware of games have also become content and context for musical production. A series of related musical practices have developed from the early practice of circumventing copyright on games. In northern Europe and Scandinavia, this community is known as the *demoscene*. The demoscene obtains its name from the demonstrations that illustrated the skills of game "crackers" who in the 1980s and 1990s "cracked" videogames by evading copyright controls or digital rights management (DRM) and thus enabled sharing of the game. Programmers sometimes spent many hours breaking the DRM and posted their nicknames on an opening screen to the cracked software to demonstrate their skill to others. Competing crackers gradually expanded simple text messages into real-time animated movies and music (Carlsson 2008), and these demo sequences themselves became the object of sharing, usually in the form of short noninteractive movie clips but sometimes as games.

Today, demo creators often try to work within the tight restrictions that were found on the early platforms (such as limiting themselves to 64 kilobytes of memory) and use obsolete or old game technology. These constraints are intentional, in part as a way to see who can push the limits of the format furthest. The demoscene group .theprodukkt, for instance, created a completely procedure-based 3D first-person shooter game called *.kkrieger* (2004) in 96 kilobytes. Many of the demo practitioners view the constraints as a challenge and as a way to learn more about programming. Some view the scene as subversive and anticommercial. Ville-Matias Heikkilä (2010) argues that *".theprodukkt* managed to critique aspects of the video game culture from inside the video game culture itself. The mere use of demoscene techniques and a tight size limitation served as a statement. . . . Personally, I like my 8-bit demos seen as having an inherent statement directed against the 'wasteful' aspects of mainstream computing—not only the wasteful use of computing resources but also to the ecological wastefulness of consumerism and high-tech snobbery."

As such quotes suggest, there is a tight community in the demoscene often marked by a closely defined politics, and to enhance this sense of community members hold regular parties and competitions among themselves. Web sites collect and share the videos, games, and music from the demos. The songs are typically produced on older MOD format sequencers (trackers) such as Fasttracker, Protracker, and Little Sound DJ, which are run on PC emulators or on the original consoles. The trackers illustrate the limitations of the technology. For example, in Little Sound DJ, songs are put together using up to four channels of sound that was originally found on the Game Boy hardware—two pulse waves, a PCM sample channel, and a noise channel (figure 4.6). Each sound is navigated by using a number input, and editing those numbers allows the player to adjust the pitch, instrument sound, and envelope (among other functions). Unlike simply picking up an instrument and making sounds, using trackers requires the musicians to take time getting a feel for the sound that is produced when they select numbers.

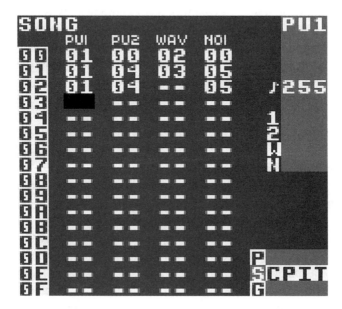

Figure 4.6
From a screen capture for a Little Sound DJ tracker run on the Visual Boy Advance emulator.

This style of music has developed into its own music genre, known as *chiptunes*. A small but dedicated group musicians makes music from otherwise obsolete computer chips and games consoles, forming bands such as 8bitpeoples, Teamtendo, and the Reverse Engineers. This can be seen as retrogaming nostalgia, as the Reverse Engineers' Edward Jones (2005) implies: "When I was a kid, I didn't listen to pop music on the radio. I listened to music on computer games. My friends and I, we would make tapes of the songs from computer games and we would listen to those. They were the tapes we'd copy and pass around." But many composers who are now in the scene are too young to have spent their childhoods with the original game machines. There are other reasons for wanting to compose chiptunes. Grethe Mitchell and Andrew Clarke (2007, 397) argue that "Their intention is simply to produce original music, and if this happens to be evocative of in-game music, it is simply because of the hardware that they use and the need to adopt certain techniques—such as using arpeggios instead of chords—due to its limitations. . . . The distinctive sound of the game hardware, particularly of handheld consoles such as the Gameboy, is attractive to musicians as it is iconic, evocative and nostalgic. It also is ideal for certain genres of music, such as upbeat dance music and 'synthpop.'" Besides the sonic aesthetic of the early game systems, one frequently cited reason for composing on old game soundchips is that composers enjoy the challenge of the limitations inherent in the technology. According to Teamtendo, "Working with this limited harmonic vocabulary forces you to be creative, and there are some very pleasant discoveries along the way" (in Katigbak 2004). As the band Goto80 says, "It's fun working with such hardcore limits, forcing you to realize your ideas in other ways" (in Carr 2002).

Punk pioneer Malcolm McLaren was involved in the scene and helped form the micromusic label. McLaren (2003) saw an anticorporate sentiment in the scene and used it to advertise the music as a new punk music: "Chip musicians plunder corporate technology and find unlikely uses for it. They make old sounds new again—without frills, a recording studio, or a major record label. It would be facile to describe the result as amateurish; it's underproduced because it feels better that way. The nature of the sound, and the equipment used to create it, is cheap. This is not music as a commodity but music as an idea." McLaren's sentiment is echoed in an article by Sebastian Tomczak (2008), who suggests that "In essence, it is the re-assignment of a device to a role that is in opposition to the purpose it was designed for that underpins this anti-consumer sentiment." But members of the chiptunes community quickly distanced themselves from

McLaren, whom they saw as fabricating a punk-style rhetoric around the genre in order to generate hype and create an artificial authenticity. An open letter to McLaren contradicts many of his claims and end with questions (Morris 2004):

> Our last point is a question that we would like you to consider. The *Observer* newspaper turns the disturbing phrase "... some of the ideas McLaren has been hatching over the past year or so, particularly his recent discovery of 'chip music,' which he thinks is the most significant new phenomenon since punk or hip hop, two earlier cultural styles which he pillaged, packaged and took credit for." The main concern of chip musicians regarding your involvement in chip music is voiced in this quote. We are wondering if it is your plan to pillage, package and take credit for chip music as well? We are uncertain of your motives—would you care to explain them?

The response to McLaren's involvement reflects the artists' fears of having the music coopted by the mainstream and watered down into something deemed more accessible. Nevertheless, chip music has enjoyed some limited underground success. *The High Voltage SID Collection* is a collection of about forty thousand songs from Commodore 64 games and new chiptunes that were produced on the Commodore 64's original SID soundchip. A few popular artists have also toyed with the chiptunes aesthetic by bringing elements of chip sounds into their music. Beck, for instance, released an EP of 8-bit remixes of some of his songs in 2005. Also popular have been 8-bit remixes of popular songs and 8-bit cover bands. The Super Madrigal Brothers, who cover mostly Baroque music, describe themselves as "electro-Elizabethan glitch-folk."

Chiptunes today are written using the original hardware and software, soundchip emulators and plug-ins for modern sequencers (such as the various Commodore 64 SID VST plug-ins), and specially written sequencing software that either emulates the original or interfaces with the original device (e.g., Little Sound DJ). As discussed above in the context of all electronically generated music, performing live can pose a problem for these musicians. As such, involving the original game hardware like the Nintendo Game Boy can allow for a more interesting live show. Here the chiptunes scene often crosses over into the circuit-bending community.

Circuit bending (also known as *hardware hacking*) is the intentional short-circuiting of small electronic devices such as sound-making toys (Speak & Spell is a popular example), synthesizers, and game consoles for the purposes of generating new sounds. The Ninbento, for instance, uses a circuit-bent Nintendo NES modification that allows a user to input music into the console and the graphics to respond to the beat. Perhaps most popular for both chiptunes and circuit-bending musicians is the Nintendo

Game Boy.[13] Game Boys can be played through custom MIDI controllers and tracker software, and Game Boy hacking guides litter the Internet. The console has become so popular among electronic musicians that there is now a small but dedicated bent Game Boy music scene. Circuit bender Reiner Zeigler explains the popularity as follows: "What makes the Game Boy attractive to the hobby designer (besides its slick look and low cost) is the great wealth of publicly available hardware and software support. Most games machines are black boxes containing custom-made hardware with little if any information on their inner workings. But a few dedicated individuals have literally taken the Game Boy apart and documented what they have found."[14] Schematics are shared online, and many circuit benders sell their creations to those who are unable or unwilling to do the work themselves (figure 4.7).

The demoscene, chiptunes, and circuit-bending communities have all been celebrated at times as antiauthoritarian or subversive. The obsolete media technologies, Garnet Hertz (2009) suggests, "reveal important themes, structures and links in the history of communication that would normally be occluded by more obvious narratives. . . . Like a time machine, artifacts

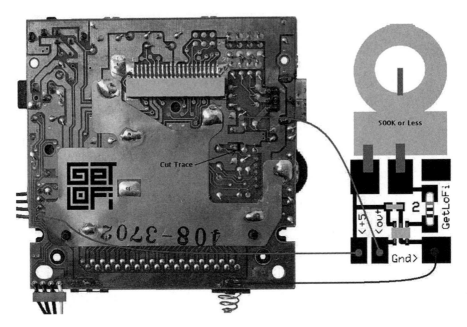

Figure 4.7
How to circuit-bend pitch on a Game Boy.
Source: Available at http://www.getlofi.com/?p=1794.

from a different era summon up a discarded mode of thought and bring forward its lost conceptual nuances. The purpose of invoking the past is to bend and short circuit the marginal past with the present: media archaeology remixes and challenges our memories of the past, the historically marginal and our experience in the present." However, many people who circuit bend or create music from the detritus of old game technology have other reasons for using the old technology. The hardware is still affordable (a Game Boy sells on eBay for about $20 to $30, although the popularity of bending them has driven up the price), so experimenting with the hardware does not cost the creator much money. People can afford to make mistakes, and this experimentation is at the heart of the bending community. Cory Arcangel (in Johnson 2006), an artist who circuit bends, argues that "Bending opens new worlds of thought, sound and composition. Bending is extremely empowering. . . . Deep experimental music has been stripped of academic trappings and is now, thankfully, spilling out into the street, a much more alive and fertile environment." Revealing the mystery of technology—by breaking open black-box technologies, discovering how they function, and reengineering them—is, as Cory Arcangel suggests, an empowering process.

The game technology of the 1980s and early 1990s was sonically unique. At that time, sound was synthesized on a chip using simple waveforms (see Collins 2008). Today's game consoles do not contain such soundchips but have far more advanced technologies that allow samples to be streamed in real time. In other words, twenty years from now, the chip community probably will not include musicians who are using the Xbox or Wii to create music. The sound of the 8- and 16-bit machines is unique, distinctive, and representative of an era in gaming and in the history of technology.

Game companies have recently tried to capitalize on fans who like to create music on handheld game consoles. The Nintendo DS, for instance, has several music programs with sound-generation oscillators, and peripherals like the Kaossilator add more modern touch-pad synthesizer technology to the devices. But as with any subculture, the demoscene, circuit-bending scene, and chiptunes scene have their its own hierarchy and definitions of authenticity. In the context of the scene, bending an instrument is more authentic than purchasing one already bent, as is writing real-time animated sequences set to music in 64 kilobytes or composing within strict limitations. For this scene, being able to do it yourself is a key signifier of authenticity because it represents a devotion and time commitment that few are willing (or able) to carry out and a skill that must be honed within the constraints of the community's expectations.

But chiptunes and circuit bending are not the only means by which game technologies are being employed as musical instruments. We have discussed two examples of modifying game engines to produce visual content for music, but there are also examples of modifying engines to produce musical content. For example, game artist Julian Oliver's *Quilted Thought Organ* (2001–2003) wedded audio samples to objects in the game modification (mod), and thus the game engine, which originally was built in *Quake II* (1997) but later was ported to *Half-Life*, can be played as a live instrument. The game engine becomes a three-dimensional music synthesizer, with sounds that the player can make by moving through objects in space and triggering the engine's collision detection. More recent incarnations of the game-instrument include *q3apd*, which Julian Oliver and Steven Pickles built using the *Quake III Arena* engine and incorporating additional means to make sound in the engine. Weapons and other objects were tied to sounds, with the player once again able to trigger the sounds by moving through the space, although the viewing angles, weapon states, local textures, player's location in space, and nonplaying character locations all influenced the sounds that were produced. The game was networked so that multiple players could play at the same time—in effect, jamming together. Oliver and Pickles (2007) have more recently created *Fijuu* and *Fijuu 2*. *Fijuu* was built in the open-source engine Nebula and allowed the player to manipulate instruments using a PlayStation controller. Six 3D instruments could be sculpted by the player to change the sound played (relying on granular synthesis, graphic filter bank, and beat pattern sequencer). Players could send short sequences to a recording ring in the center of the gamespace, which could be layered and rotated on this ring, producing a composition (figure 4.8).

Although modified game software allows for new musical interface devices, the hardware itself can also become a new means to create music. Nintendo encouraged the use of Wii as a musical instrument through games like *Wii Music* (2008), which are not merely musical play-back games like *Guitar Hero* (2005) but allow the player to alter and construct songs. Although *Wii Music* has fifty selected songs, players can alter a song's tempo, remove or add notes, and change the emphasis of the beat to modify a well-known popular song into something unrecognizable and new.

The popular use of Wii remotes as a musical instrument in nongaming practice suggests several parallels between gaming and musical instrument playing: "If you're an electronic musician with a hankering for something new, the fun really starts when you add a Wiimote," writes one journalist

Figure 4.8
Fijuu.
Source: Image from Peter Kirn (2006).

(Lehrman 2009). In describing the band tokoleten, which uses a Wii remote as a sound controller, music journalist Peter Kirn (2009) notes that "It's proof that the controller—any controller—is in the hands of the creator, and what it sounds like is entirely undetermined. . . . Controllers are always abstracted from the sound, by definition, and whether they're satisfying to you depends on how you've mapped them. I don't know what qualifies as innovative, but then, there have been times when I've very much enjoyed turning a knob, so 'innovation' isn't always what matters to me. I tend to fall back on Duke Ellington—'If it sounds good, it is good.' For controllers, that means 'If it feels good, it is good.' You're the one with the controller in your hands." The key to music-based controllers seems to lie in this "feel good" tactility and in their gestural congruence, which brings an embodied connection back into electronic music production. This ability to mimic acoustic instruments with an electronic instrument enables the role-playing and interaction with the game that facilitates the game as performance.

Interacting with the Game as Instrument

Roland Barthes (1991, 188) wrote that "the 'grain' [of music] is the body in the voice as it sings, the hand as it writes, the limb as it performs." As discussed in chapter 3, we are connected to a performer of a piece of music through our own vicarious performance of that music and through our own embodied experience of sound making. The seeming need to pair music with visuals comes from a similar need to "materialize" the music and give it body in the face of its digital disembodiedness. This materialization, however, is not just a desire for a visual accompaniment to music: it is also a desire for a corporeal connection to that music. In part, the immateriality of much game music—the lack of grain—encourages people to play with the sound. By making music through an interaction with game technologies, players can make the gameplay experience tangible, removing it from the ephemeral nature of the game and imbuing it with a sense of aura.

The intentional use of devices that introduce chance and accident into digital performance, such as circuit bending and live-streaming performance, similarly recalls a need to introduce a more organic, human, and live element into the music. Circuits age and sound different as time passes, creating a "living instrument" that is altered every time that the machine is turned on (Ghazala 2004, 101). Thus, the imperfections of the system can "be said to constitute a form of sonic 'grain': a 'space of encounter' between music and 'noise'—embodied and disembodied sounds—whereby the latter can become aestheticized as a valued component of the listening experience" (Wallach 2003, 43). By introducing the chance that is inherent in activities like playing game engines as instruments or circuit-bending hardware, the performers signify liveness to an audience.

The sonic appropriation from video games can be viewed as a form of found sound, a practice that is as old as music making itself. The use of games as material can be considered to be part of a long tradition of folk handiwork that uses everyday objects to create something new. Video games are a modern version of that detritus that forms part of our cultural soundscape, whether the hardware or the software, and thus becomes another material with which to work. They have become part of a tradition of musical practice in which "each soundscape composition emerges out of its own context in place and time, culturally, politically, socially, environmentally and is presented in a new and often entirely different context" (Westercamp 2002, 52). For many of us, the sounds of video games are a significant part of our daily life. Cory Arcangel (in Houbt 2004) explains,

I'm 25 and I have no experience with anything except media, so it's like, I can't make anything. . . . The language I understand is media, so when I make something, as a raw material it's the only thing I'm comfortable with. It's not a conscious effort, being a hacker or making a political statement. . . . It doesn't make sense for me to make work out of anything else. It doesn't make sense for me to just draw stuff. I think with a lot of artists my age, it's all just mashing stuff together, and it's all about connotation and it's all about how things fit together, and it's all about cultural references.

The games themselves have become the media through which music is expressed. They provide the visual context of performative expression in the case of virtual performances, the lyrical content in the case of covers and filk songs, and the instruments in the case of chiptunes and software and hardware hacking. This use of games as musical instrument is leading to new innovations in nongaming musical practice. For example, after the success of Wii controllers as musical devices, Yamaha created what it called the Muro sensor, an accelerometer-based wireless device that uses nearly the same technology as the Wii remote. Such devices have brought game-based music beyond the game-playing audience to a wider musical base and afforded new means for embodied interaction with electronic music.

But the interactivity of game sound encourages users to continue that interaction beyond what was originally planned by game developers. Many of these types of performative activities indicate a strong desire in players to go beyond the standard means of player-generated content. Often, this content is unintended by the designers, but it helps to increase their product's lifespan. Game Boys are still selling well as used products since being given new life through their use as a circuit-bent musical instrument. This type of interactivity in part represents the desire of players to personalize games and make products their own. But such interactivity also alters the original meaning of the games: game designers may resist opening up their games to such activities because they lose control over the content. This experience of customization and cocreativity is explored in the final chapter.

5 The Second Life of Game Sound: Playing with the Game

The practices described in the previous chapter illustrate some of the ways in which players extend the experience of the game beyond game play. Artist Julian Oliver (2006), who modifies game engines into musical instruments, asks, "Is artistic modding an abuse of the game? Yes of course. But it is not an abuse of the medium. To abuse the *medium* of the game is to merely play it." Oliver suggests that modifying games beyond what they were intended for is inherent in the technology and is not only encouraged but demanded by that technology. The shift from the "read" to the "read/write" culture (Lessig 2008) has encouraged a wide range of cultural practices, but digital interactive media like games have advanced this practice because they are able to alter, coopt, and adjust content (both hardware and software, as seen in the previous chapter) relatively quickly. Science fiction writer William Gibson (2005) argues that, "Our culture no longer bothers to use words like *appropriation* or *borrowing* to describe those very activities. Today's audience isn't listening at all—it's participating. Indeed, *audience* is as antique a term as *record*, the one archaically passive, the other archaically physical. The record, not the remix, is the anomaly today. The remix is the very nature of the digital. And if *audience* is an antique term, then equally so must be the concept of the author."

Notions of cocreativity and user-centered design have permeated many products and services that are used today, from low-tech running shoes to high-tech mobile phones. The ability to customize a product allows users to express themselves, to have a sense of agency and ownership, to feel in control, to accommodate emotional states, and to have fun (Mugge, Schifferstein, and Schoormans 2010). Digital technology like video games has more recently allowed for wide-scale customization. On the Internet, for example, the notion of Web 2.0 and user-driven content has grown in a single decade to the point where it is now taken for granted. Participation—as a form of interaction—is one of the hallmarks of new media, and this includes cocreative practices.

This blurring of the line between author/creator and reader has become a focus of cultural studies in the last few decades, and scholars like Henry Jenkins (1992, 2006a) have shed light on the many meanings of texts and also the new texts that fans produce from their involvement with media. Jenkins has written extensively about the creation of new meanings and products from texts in terms of active reception, interpretation, consumer action, and cultural production. In his model, an audience may "poach" materials from media and use them as a basis for their own social community (Jenkins 1992). As with John Fiske (1992), Jenkins believes that although all audiences take part in the production of meanings, fans take this a step further into the production of their own forms of texts.

The terms *customization* and *personalization* are often used interchangeably, and writers often disagree about their definitions. Here I define them as follows. *Customization* is built into the game by the game creator so that players can select certain already designed features. Some games, for instance, are designed to allow players to create their own custom avatar. Settings in the menu screen allow players to select whether they want to hear sound effects, ambience, music, and dialog. This sonic customization takes place according to a predetermined series of features or options that are built into the game. *Personalization* is not planned by the designers but is created or hacked by end users. In gaming, these can be hardware hacks (modification of the game controls or hardware), software hacks (modification of the game), remixes, and creations that arise from using game content in nongame settings. In simple terms, player alterations can be intended (customized) or unintended (personalized) by the game developers. There is a fine line between the two, however, because when products become personalized by enough people, the personalization practice is often coopted by the developers and turned into a customization feature for future versions.

The cocreative practices described in the previous chapter developed from out-of-the-game content, game hardware (game consoles), and game engines (core software that houses the game's design). This chapter focuses on the sonic customization and personalization of game engines. Game engines are often used by the industry to save on the costs of reprogramming standard elements and built-in functions such as collision detection, rendering, and artificial intelligence. Here I describe how creators repurpose game engines for their own activities. In this way, the processes are not always the developer-endorsed practices of customized modifications but instead are unsanctioned (and often illegal) software hacking. The term *hacker* has often been used by the popular press to describe software programmers who illegally circumvent a security system, but the concept of hacking is much wider (and much less threatening). *Hacking* can be more broadly defined as "interacting with a computer or any other technology-infused system in a playful or exploratory way, or modifying an existing system (hardware, mechanical, or software) to improve performance or create an application that differs from the device's original purpose. . . . The true hacker is an individual who can achieve miracles by appropriating, modifying, or 'kludging' existing resources (devices, hardware, software, or anything within reach) to suit other purposes, often in an ingenious fashion" (Paradiso, Heidemann, and Zimmerman 2008, 13).

This chapter investigates the hacker aesthetic in the game world and the sonic creations that result. I describe the many unsanctioned ways that players rework or remix video game material and prolong the life of game material, which goes on to exist in new ways long after the games are no longer available in stores. As Jonathan Lethem (2007) states, "In the first life of creative property, if the creator is lucky, the content is sold. After the commercial life has ended, our tradition supports a second life as well." For the most part, this second life lies outside the sanctioned purview of the official game industry, and in some cases it belongs to a small independent and underground economy of a sort. Often this work is given away for free—sent back out into the world for further hacking and remixing. A recursive postproduction practice takes place as creators comment on the work of other creators. In this way, the original game, the new work, and the creator become a part of the interactive process. Game sound thus becomes an interaction between player and game, between players, and between player and society (through comments on culture or on the games themselves).

Here, I examine game sound personalization and customization as forms of interactivity and look at the effects that customization and personalization might have on a player's experience. How do players customize and personalize game sound and why? Game soundtracks are carefully composed around a game's emotional content, genre, style, action, and narrative. With the ability to customize music for games, what happens to the experiences of the player when the music is changed? How might customization and personalization affect players' identification with game characters?

Sonic Modification and Player-Generated Content

As a component of social interaction, player-generated content is one of the driving factors of online games. *Player-generated content* in this context can be defined as the objects, actions sounds, and events that occur in a virtual space that are contributed by players and that are not predefined (that is, preprogrammed or prescripted) by the designers. For example, players can create objects and upload them so that other users can use them in their own games. In *The Wealth of Networks*, Yochai Benkler (2006, 74) describes the function of massively multiplayer games as places "to build tools with which users collaborate to tell a story. . . . [Players] produce a discrete element of 'content' that was in the past dominated by centralized professional production. . . . This function is produced by using the appropriate software platform to allow the story to be written by the many users as they experience it." This distributed authorship distinguishes MMOs and online virtual worlds from most other types of video game play: the story is not scripted or created by developers in advance but unfolds over time through the creative and performative practices of the players. This is not a choose-your-own adventure but a create-your-own adventure. Axel Bruns (2007) argues that in the context of virtual worlds the creative component is so integral to play that current terminology is inadequate: "the very idea of content *production* may need to be challenged: the description of a new hybrid form of simultaneous production and usage, or *produsage*, may provide a more workable model." Another way to describe such media is cocreative (Morris 2003): neither developers nor players are the sole creators and mediators of a game, but through the personalized act of play, players bring their own content, meanings, and ideas into the game. Shared meanings are created in the virtual space, and these stories are coconstructed and told between groups of people who all contribute content.

  EverQuest, launched in March 1999, was one of the most popular early online massively multiplayer games, with nearly half a million subscribers at a time when the Internet was far less ubiquitous. Although auditory latency was common in those days, *EverQuest* cleverly combated both latency issues and repetitiveness by introducing the concept of allowing users to tie custom sound effects from their own computers to events whenever a specific phrase appeared in the text-based chat window, a function that was part of a late 2005 update called "Audio Triggers." As text messages scrolled across screens, players needed to react quickly to certain phrases and could be alerted with an auditory warning that was tied to keywords. For example, if players wanted to know that they have just been kicked, they could set an alert for "kicks YOU." If the phrase was "A frost giant savage kicks YOU for 30 points of damage," players immediately knew that they had been attacked, before they could even read the phrase. Sound files were stored on the player's server, and only the player heard the sound, unless players were using voice chat and the sound played loudly enough to be heard through the player's microphone. In such cases or when players told other players what triggers they were using, other players occasionally sonically "spammed" that player by repeatedly typing in the phrase to trigger the audio file. On an *EverQuest* forum, a player explains, "Never tell guild mates your audio triggers or you will get spammed. . . . A warrior in my last guild had 'enrage' set as his audio trigger and the audio was of a gong sound. Several of us found this out and in between fights we would /tell Vortimer 'enrage' and listen to his bongs whenever he spoke on Ventrilo."[1] Despite the risks of sharing trigger keywords with other players, players liked to discuss their unique approaches to using sound effects on Web sites and in-game chat, often sharing ideas and trying to outsmart others by attempting the most witty or creative uses of sound effects.

Another example of allowing player-generated sound effects to be tied to objects or events in the virtual world can be found in *Second Life* (2003), in which players can tie sound-effect samples to objects that they have created. For example, a player-generated motorcycle can play motorcycle sound effects (or other sounds) that are selected or produced by the creator. When creating the object, the creator sets permissions that allow others to modify that object by adjusting sounds, creating new sounds, and so on (see Marcus 2007). Players can sonically customize objects in the world and can add humor or their own style to the overall soundscape of the space. In this way, players bring the real world into the virtual (and vice versa) by acknowledging the artifice of the creative practice and intentionally bringing that artifice into the game. The sound effects may be completely unrealistic or unconnected to the object/phrase involved, yet the act of tying them to each other and sharing them with others extends the real world into the virtual space. As players select their own sounds and tie these sounds to particular events or objects, they are sonically creating their own personal virtual space, a further step toward seeing the virtual space as an extension of the real. Just as we have a different relationship to self-produced sounds (see chapter 2), it may be the case that we also have a different relationship to self-selected sounds. In this way, this type of interactivity with sound may likewise facilitate identification with the characters and immersion in the space.

In addition to sound effects, player-generated musical content has also been incorporated into some games. *Grand Theft Auto* (1997), *The Sims* series (2000, 2004, 2009), and various *Gran Turismo* games (1997, 2010) have made concessions in the design to accommodate player music. In the *Sims* games, for instance, players can swap out the game's preselected music and play their own once they have purchased a stereo system. In *Grand Theft Auto*, players can play their own music through car stereos. For Kiri Miller (2007, 404), "The player-controlled radio stations not only increase the verisimilitude and immersive qualities of each gameworld, but also encourage players to associate particular music with particular characters and places" (see below). More commonly than having such affordances built into games, however, are the many modification practices that allow players to bring in their own sonic content.

Modding Game Sound

The advent of computer game modifications (mods) has been one aspect of the drive toward user-customizable game components. In her dissertation on the modding culture of *The Sims*, Tanja Sihvonen (2009, 59) defines *modding* as "the activity of creating and adding of custom-created content, *mods*, short for *modifications*, by players to existing (commercial) computer games. These additions can be supplementary—in which case the mod is called a *partial conversion*—or mods can result in an entirely new game, which is then called a *total conversion*." Modding includes several activities that can overlap and that depend on the abilities and desires of the person (or persons) modding the game. Modding can include simple changes in the graphics or sound of weapons or characters, more advanced mapping, and entirely new versions of a game based on the original game's engine. In the previous chapter, an example of game modding for sonic purposes involved the use of game code as background visuals to music. Here, I explore other sonic modding practices.

Modding has a long history of practice in the game industry. *Castle Smurfenstein* (1983) was perhaps the first. It was an early modification (and parody) of the Nazi-shooter game *Castle Wolfenstein* (1981) and was written for the Commodore 64 and Apple II (figure 5.1). According to the authors, "The Nazi guards became Smurfs, the mostly unintelligible German voices became mostly unintelligible Smurf voices. We created a new title screen, new ending screen, new opening narration, and an opening theme, and changed the setting from Germany to Canada" (Johnson 1996). Modding became more common practice with *Doom* (1993). The developers, id Software, published the source code of the game in 1997, and level editors were designed by players, a move that has been described as a "watershed in the evolution of the participatory culture of mod making. Anyone with

Figure 5.1
Castle Wolfenstein (1981) and mod game *Castle Smurfenstein* (1983) (image from Johnson 1996).

the interest could create a level of a complex game, the equivalent of writing a new chapter into a book, and then, via the Internet, publishing that creation" (Kushner 2002, 71). The success of the *Doom* modding community led developer id to release future games, such as *Quake* (1996), with open source code. *Quake* led to more widespread modding practices, in part due to the simultaneous rise of the Internet. Since that time, modification communities or modders have become a significant marketing factor for computer game developers, especially in first-person shooter, role-playing, and real-time strategy games.

In addition to satires, parodies, and their own creations, game fans have also re-created games from the past that have become difficult to obtain and play on modern machines. Anastasia Marie Salter (2009) documents how fans have carefully recreated 1980s adventure games and how legal battles have ensued. Although some reconstructions attempt to recreate the original game with updated graphics, voice acting, music, and other technology (such as the re-creations of the *King's Quest* series by AGD Interactive and Infamous Adventures), many fan sequels keep the original characters and settings but add a new narrative and new puzzles. The LucasArts game *Zak McKracken and the Alien Mindbenders* (1988) has been repeatedly extended (for instance, as *The New Adventures of Zak McKracken*). Fans have even made their own software engines to facilitate the reconstruction of old game genres, with or without the original characters or storylines. In this way, the players interact with other players and with the original creators. Salter (2009) notes that

The fan author is engaging in a one-way dialogue with the works of the previous creators: the fan is remaking the classic game, and the original creator is now silent beyond their original production. This is perhaps best understood as a practice that extends [Espen J.] Aarseth's consideration of the adventure game genre as folk art, as referenced earlier: works are put into the communal tradition, and new works emerge that continue and expand upon that tradition ([*Cybertext: Perspectives on Ergodic Literature*] 1997, 100). Who is the ultimate author of the work? All the creators involved in the practice. There need be no notion of one auteur, of one author working alone to create a masterpiece.

Such reuse of intellectual property—either to recreate the original game or to use elements from the game as components of a new game—occurs when players find new ways to extend their enjoyment of the game. Whether this is the re-creation of a specific game or a new creation in a particular genre, players are filling a need that the industry has failed to meet. Unfortunately, game companies have forbidden many of these creations under the guise of copyright infringement, which has marginalized the practice. As with modding, the game companies could capitalize on some of the creativity that players provide and assist these smaller niche markets.

Modders spend considerable time editing a game's code, sound, and graphics to develop a portfolio that will enable them to obtain a job in the industry, for social or cultural capital, and "for fun or out of love for a particular community or game" (Postigo 2010). This sense of community, argues Postigo, is a key factor in players' involvement in modding. Many mods are a way of showing off prowess and skill at modding. As illustrated in the *Smurfenstein* example, mods are also commonly used for satire, spoofs, and other modes of social commentary or self-expression. The sense of community and the expressive elements of modding have led academics to situate modding culture as a fan culture (see, e.g., Sotamaa 2004). This distinction sometimes sits uneasily. Although "One of the traditional claims of the fan critics is that fan cultural texts are not produced to make profit," sometimes mods are created with the hopes of some profit (Sotamaa 2007, 113). There have been a number of notable financially successful cases, such as *Counter-Strike* (1999), a mod of *Half-Life* (1998).

The fine line between fans and the corporate world has led some to criticize the game industry as taking advantage of the free labor of fans to minimize labor costs, extend shelf life, create brand loyalty, and reduce research and development and training costs. Up to 90 percent of *The Sims* content, for instance, is said to have been produced by players (Sihvonen 2009). In an interview, *Sims* and *Spore* (2008) designer Will Wright says, "I didn't want to make players feel like Luke Skywalker or Frodo Baggins. I wanted them to be like George Lucas or J.R.R. Tolkien" (Borland 2006). It may be the intent of the game designers to encourage the role of the player as producer in these games and allow for creativity, but this also reduces the amount of content that the creators need to develop. Describing the practice as *playbour* (play labor), Julian Kücklich (2005) maintains that the industry "benefits from a perception that everything to do with digital games is a form of play, and therefore a voluntary, non-profit-oriented activity." Sanctioned mods (that is, mods that use game engines that allow for modding) typically remain the property of the original game developer, leaving modders without ownership over their creations, while the games industry exploits the free labor and reduction of risk that are associated with player-generated content.

The practice of modding sound includes various modding activities. As noted above, some game companies are open to the practice of modding and release source code to encourage fans to edit and share new levels or content based on the game. In these cases, the entire game can be redesigned, and new sounds can be incorporated into the game based on the altered needs of the modification (such as Smurf sounds instead of Nazi soldiers). Some games are not designed for modification, in that the developers do not release the source code. But according to an unwritten rule that is assumed for many PC games, access to the library of game assets (such as sound files) is possible, and the player can overwrite the original sound files by copying in new files with the same names, without changing the actual game in other ways. Game Web sites are set up by players or game designers to help users identify these files. Moreover, some players share soundpacks of sound files that they have created for the game for the purpose of substituting files. Some players develop modding utilities that aid the user in automatically overwriting music and sound effects. For example, the audiomod for the independent game *Minecraft* (2009) allows the user to add new music and sound effects to the game alongside the originals. *Soundtrack* is a mod customization plug-in for *World of Warcraft* (2004) that allows players to swap out the game music with the players' own MP3s (figure 5.2). The mod advertises: "Ever get tired of *World of Warcraft*'s default music? This *World of Warcraft* soundtrack plugin allows you to customize music for *WoW* zones, combat and dance, using a very easy to understand interface. Wish you could use your *L.O.T.R.* music while in Elwynn Forrest? *Zelda* music when you engage in PvP? *Bootilicious* when your troll girl starts dancing?"[2] With this particular mod, the player can assign music to specific zones or subzones, monsters, or events. This may help to reduce some of the disassociation between music and game that can occur when the music fails to synchronize to the action of the game (see Wharton and Collins 2011).

Figure 5.2
Soundtrack, a music substitution mod for *World of Warcraft* (2004). The image shows songs from other games, such as *Lord of the Rings* and *Metal Gear Solid* (1998), that will be substituted for the game's predefined music.
Source: World of Warcraft Mods: Soundtrack 1.8, available at http://www.warcraft -mods.com/Soundtrack.html.

Game players enjoy sharing these soundpacks of music and sound effects and have created YouTube videos of some of their modded games to demonstrate their skills. For instance, a modification of *Half-Life 2* (2004) using voice-generated sound effects on YouTube has received nearly one and a half million views and listens.[3] Instructions, software, and the code that allows users to create their own mods are also commonly shared online, such as *Dance Dance Revolution* (1998) hardware and software hacks that allow users to input or compose their own new music for the game (Höysniemi 2006). A similar sharing of software to modify game music has also been created for *Guitar Hero* (2005), in which versions of the game have a variety of mods that enable players to create their own note tracks and import those songs into the game. These custom tracks are often shared online on Web sites like Scorehero,[4] extending the gameplay well beyond the limited selection that comes with the game. In fact, the idea was later coopted by the game's developers, who formed The Rock Band Network, which allows players to score their own songs and share those songs with other users for playing in the game.[5] This reinforces Hanna Wirman's (2009) contention that game modding cannot be considered purely a resistant activity and that the motivation for modding "may derive from a wish to continue one's experiences with a particular game even longer and in new ways."

The player's ability to modify music has also been built into a few games where designers have found ways to allow the player choice while maintaining some control over the auditory content. In *Grand Theft Auto: San Andreas* (2004), for instance, players can select stations on the radio of the car that they are driving. Here, the designers have found ways to incorporate player control by tying it to diegetic music in the game, thus reducing the effect of having the music take the player out of the immersive experience. Kiri Miller (2012) writes about the player's ability to select (predetermined) songs in *Grand Theft Auto: San Andreas* and notes that the players that she studied often chose to listen to music as if they were the main character—that is, through their character's ears. Many players chose music that they thought their character might listen to (mostly hip-hop) rather than music that they wanted to listen to and likewise selected music for particular moods. She discovered that players developed a taste for different music and listened to music in a new way. Familiar songs took on new meaning in the context of the game. This suggests that games can become (and are becoming) a new way of listening to music in general. Whereas the music video brought in new ways of listening to music (through watching a narrative or performance video), video games are creating a new way of listening through in-game interaction.

Likewise, Wharton and Collins (2011) found that altering the music in a game changed meanings, actions, effects, and emotional response to the game. By varying the songs or altering the order of the songs, players reported different immersive and emotional states and also considerably changed the ways in which they played the game. New meanings were created through juxtaposition and counterpoint of music and game. At times, songs that were chosen took on an irony by being juxtaposed with the violence of the game. In other words, the overall semiotic meanings of the game can change considerably from what the game's designers intended. The players in the study consciously or subconsciously attempted to make connections between the music that they chose and the game's narrative, events, imagery, and playing tactics. Players found coincidences between elements of the music and actions on-screen and chose music that they felt would increase their enjoyment of the game.

Although customization may be empowering for the players, at stake for the developers is control over their intellectual property. For game designers, the trend has meant a step backward in terms of the idea that their music is an essential part of the entire *gesamtkunstwerk* of the game. Customization might be a desired trait for the consumer, but often the designers of the game (along with the composers) must relinquish control over the musical soundtrack to the game. Such a feature would be unheard of in film, says sound director Rob Bridgett (2010, 23), who notes that "One cannot imagine, for example, removing Howard Shore's score from such a fully integrated work as *The Lord of the Rings* movies and replacing it with user defined content." Unlike a film, he adds, games offer individualized play: every game is different, every player is different, and the customization of music can be seen as another layer to this individualization, perhaps as part of the larger modding concept of games. In other words, interactivity itself can be seen as a form of customization.

The ability of the audience to alter media content is an essential component of interactive media and has long been celebrated as democratizing and empowering by cultural theorists. John Fiske (1992) argues that fans (that is, "productive" users of media) are often in a resistant relationship to the commercial media industries and through their fandom create an alternative cultural industry with its own production and distribution systems that lie outside the mainstream industry. Likewise, other authors have celebrated the democratizing potential of user participation in media (see Banks and Deuze 2009). However, the concept of the cocreator is misleading in that it suggests an empowerment of the user that may not exist. The potential of being simultaneously a consumer and a producer is not the equivalent to being empowered, since the right to distribute has become even more important than that to produce or reproduce content (Kücklich 2005).

However, many authors have pointed out that modding exists in a different position to other democratized media practice. First, modding is often accepted and encouraged by the game companies, providing the game companies with the benefit of longer shelf lives, stronger brands, and a sense of community for players, and second, modding can lead to monetary gain on the part of the modder. Moreover, modders must work under strict licensing: often that leads to a significant advantage on the part of the game company at the expense of the modders (Sihvonen 2009, 158). It is not unreasonable to view the practice at least in part as exploitation. After all, "one of the main objectives of the games industry is to make sure that the player does not reflect on these forces" (Kline, Dyer-Witheford, and de Peuter 2003, 19). Milner (2009, 494) takes an equally cynical viewpoint, arguing that the position of game companies are somewhere between "suppression and supervision," with the ultimate end goal of higher profits and a stronger brand. However, this exploitation should be viewed as strictly a commercial and financial exploitation, since modders still obtain enjoyment and cultural capital from the practice.

The ambiguity and mixed signals from game producers toward modding often leads to emotionally charged confrontations in which "modders find themselves frustrated because of their inability to creatively work with the content they love. Furthermore, their supporters and game fans in general are angered because they cannot access innovative mods" (Postigo 2008, 61). And although the industry-sanctioned reworking of video games through modding and other practices have brought forth some tricky legal battles, the remix culture that has developed around games—very little of which receives approval or support from the industry—is even more mired in copyright questions but is a much wider practice.

Some of the legal battles that have ensued over player-generated content have arisen around the concept of the game as a brand. In this way, "the interest in the integrity of the characters is not an interest in market share, but a general reputational concern, which copyright law does not formally recognize" (Tushnet in Taylor 2006, 144). For instance, Marvel Comics sued the developers of the game *City of Heroes* (2004) because players had the ability to customize content to make their characters look like Marvel superheroes. Square Enix filed a cease and desist order against one fan mod of an older game's ROM, or game image file, *Chrono Trigger* (1995), called *Chrono Trigger: Crimson Echoes*. Fans worked on the game for five years before releasing a trailer that caught the eye of the original developers. Similarly, other fan productions like *World of Starcraft* (Chalk 2011) and fan game *Nexus: Battlestar Gallactica* were shut down. One mod called *Duke It Out in Quake* used the (copyrighted) *Duke Nukem* (1991) character in the 3D *Quake* (1996) engine—mashing two companies' products together. Despite attempts to stifle the mod, fans downloaded and played the game on the Internet, ignoring copyright laws, which Postigo situates within Henry Jenkins's (2006a) concept of the "moral economy" in which fans justify when the appropriation of content is acceptable.

The arguments made against such mods are that player-generated content may be of poor quality, offensive, or illegal; that the reputation of the copyright owners is at stake, since offensive material may alienate some

players; and that player-generated content that infringes on other companies' copyright can lead to legal liability issues for the company (Lastowka
2008, 912). There is a tradeoff between allowing players the freedom to be
expressive and creative and maintaining control over intellectual property.
Indeed, a sampling of ROM hacks available at Badderhacks presents us
with "Super Addict Bros," "Retard City Rumble 3: Tards in Time," "Skin-
head Fighter," "Super Vietnamese Hooker Bros," and other material of
questionable taste. But Lawrence Lessig (2004, 9) argues that far from being
concerned about intellectual property, "This is not a protectionism to
protect artists. It is instead a protectionism to protect certain forms of
business." Allowing any precedent to be set that may enable players to
reuse intellectual property for their own means and gains, it is felt, will
open up the practice to widespread infringement. But some modding
practices—art games—have (until now) remained largely outside the copyright and intellectual property skirmishes of fan mods.

Art Mods

Video games have also become a somewhat accepted part of the art world
as source material, subject, and tool. Exhibitions about video games are
proliferating (such as the one held at the Smithsonian American Art
Museum in 2012), and write-ups on game art regularly find their way into
art magazines and journals. Michael J. Thomas (1988) traces the history of
game-based "playful" art to Marcel Duchamp and Öyvind Fahlström, but
the use of video games as an art form did not begin until video games
themselves became mainstream. Although some games, such as Toshio Iwaii's
Electroplankton (2005), have questioned the nature of art versus game, distinctions can typically be made with the marketing and distribution of the
creations (art exhibition versus retail) as well as the intent (commentary
versus commercial success). These are not strict divisions, and unless the
creators themselves declare that the product is art, many works will sit on
the fence between art and game.

Art mods use an original game ROM that is modified in some way
by the artist and that may "employ game media attributes, such as game
engines, maps, code, hardware, interfaces etc, for a very broad range
of artistic expressions—abstract, formal and narrative, as well as cultural,
political and social" (Cannon 2003). As artist Julian Oliver (2006) argues,

art-mods innately challenge a mass-market-driven design paradigm where consumers are gathered into large interest groups or demographics and then marketed games
on that basis; every mod can be seen as a way of personalising the original and, in

so doing, they affirm that the medium is larger than the form in which it was given, that it is larger than the game. For this reason artistic modding is especially historically important; it steps completely out of the market of intended use, yet becomes intimate with the product as a material level. It peels back the layers of awe, techno-prowess and glitz to reveal it as a system of working parts that can be understood and repurposed.

A ROM is the (software) code from the game, and typically it is played back on an emulator, which emulates the original machine. ROMs are readily available online, and when the code is opened up, the program can be edited to cause fundamental changes. ROM hacks can "take any number of forms, from game-specific editor programs to manipulate or view certain variables, art content or level designs, to full-fledged content extraction from a ROM for dissemination and use in general-purpose viewers, editors and players" (Jordan 2007, 712). For example, Cory Arcangel's hack of *Super Mario Bros.* (1985) into *Super Mario Clouds* (2002) erased all the objects from the game except for the clouds drifting by. *Gameboy_ ultraF_UK* (2001–2002) by Gavin Corby and Tom Baily uses a modified Game Boy emulator that slowly degenerates as the player plays a game, making the game more and more unplayable over time (figure 5.3). The more the player plays the game, the more impossible that play becomes. Andy Clarke and Grethe Mitchell (2007, 17) suggest that "These artworks show how it is possible for a piece to comment intently upon the nature of games without actually being a game or—more accurately—by frustrating the user's expectations of what a game should be and how it should act."

Of particular interest here are the music-based art mods that have been developed, such as the short-lived trend of Automatic Mario.[6] Automatic Mario mods were hacked *Super Mario* ROMs, which required little or no input from the player, generating a predetermined movie sequence that was similar to demo video superplays (demonstrations of skill mastery in the game). The Automatic Mario hacks are always set to music, often made by using a custom game ROM hack that incorporated sound blocks that Mario could hit to make sound, adapted from the Mario Sequencer software that came with the *Mario Paint* (1992) game (figure 5.4). A Windows version of the sequencer was created and released, making it easier for Automatic Mario creators to include custom sound blocks and bumper blocks in the hacks that propel Mario through the level.

An especially elaborate Automatic Mario was set to Queen's "Don't Stop Me Now," with four separate screens of the *Super Mario World* (1990) game occurring at the same time, each one representing one of the members of the band, synchronizing sound effects and action to the music (figure 5.5).

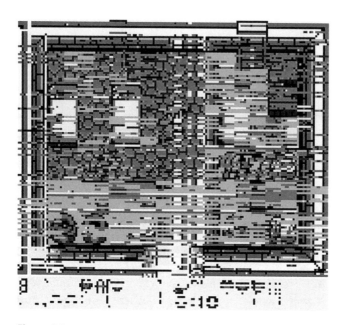

Figure 5.3

Gameboy_ultraF_uk (2001–2002) by Gavin Baily and Tom Corby uses code as ready-made.

Source: Image from *Reconnoitre,* available at http://www.reconnoitre.net/gameboy/index.php.

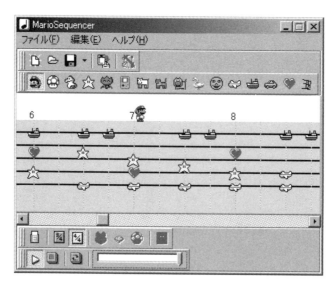

Figure 5.4

Mario Sequencer, with sound effects from *Super Mario Bros.* (1985).

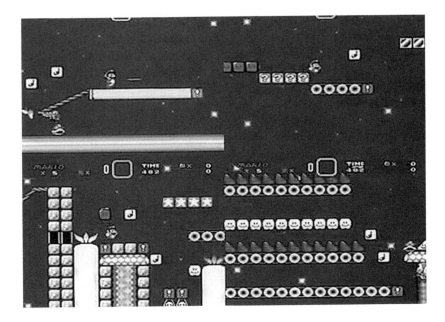

Figure 5.5
Automatic Mario Queen game.
Source: From Joystick Division, available at http://www.joystickdivision.com/2009/11/automated_mario_queen.php.

In this case, the automated portion was made into a video and shared online, although most Automatic Marios are not videos but actual software that the player can play over an emulator. In this way, Automatic Mario stands out as a ROM hack, as it removes the player from the action: the game plays itself, and thus the interactive connection between player and game is lost.

Such removal of the player calls to mind Walter Langelaar's *nOtbOt* (2007), an automated game that controls—and is controlled by—a first-person shooter game in a feedback loop. The view angle is generated by a virtual player and is looped back into the mechanized joystick. Artist Julian Oliver said of the piece, "Walter Langelaar's *nOtbOt* antagonises conventions of games as slave to our control. *nOtbOt* decouples the user-agent from the input chain, leaving just a joystick thrashing about in response to every twist and turn of a bot rampaging through a stock *Quake III* level. My first impression of *nOtbOt* was of a haunting: an AI that would take no more, fighting back at the input device in an urgent attempt to disenfranchise itself from a history of bondage."[7]

The removal of the body from play is interesting in that technology that is designed to be interactive has had its interactivity removed. These hacks are used to comment on the nature of interactivity and games. Automatic Mario is part of a wider commentary on gameplay and the body that includes, for example, Kaizo Mario ("Asshole Mario") and other Mario AI games where gameplay is taken to an extreme that is nearly impossible for the player to accomplish. The games become a display of technical virtuosity, but the game performer, in a sense, becomes the artificial intelligence rather than the human being and repositions the player as a noninteractive spectator. This emphasizes our own inability to match the skills of the machine: our real body fails to live up to the virtual body of the game engine.

Player-Generated Content: A Fourth Wall of Sound

This chapter has demonstrated how games can become mediators of social interaction and the site of various performative and cocreative activities beyond the original context of the game. The cocreative and customization practices described here illustrate some of the ways in which players of games can extend their enjoyment of the game. Through adding to or modifying game content, players can bring their own personality, style, and preferences into the game and thereby both play the game and play with the game. The creative work that is discussed in this chapter can also be viewed as a form of *détournement* in which objects or creations are used in opposition to the original meaning. As the postpunk appropriationist pioneers Negativland (n.d., 92) describe: "We are now all immersed in an ever-growing media environment—an environment just as real and just as affecting as the natural one from which it somehow sprang. Today we are surrounded by canned ideas, images, music and text. . . . The act of appropriating from this kind of media assault represents a kind of liberation from our status as helpless sponges which is so desired by the advertisers who pay for it all." In other words, the practices can be seen as an attempt to talk back at all the media bombardment that we experience, and interactive media that already enable customization are perhaps the ideal media for that talk-back. Katie Salen (2002) situates these types of practices as potentially resistive, referring to them as a form of transformative play: "Because the creators of emergent systems, like generative music or games, can never fully anticipate how the rules will play out, they are limited to the design of the formal structures that go on to produce patterns of events. Sometimes the forms of play that emerge from these

structures overwhelm and transform, generating rich and resistant outcomes. Sometimes, in fact, the force of play is so powerful that it can
change the rule structure itself." Whether or not some form of resistive
intent is behind these types of activities, they nevertheless can be viewed
as a type of game play and cocreativity inspired by the technology and by
the dominance of games in our social lives today.

In addition to breaking down the barrier between players and creators,
cocreative practices are also critical to breaking down the barrier between
audience (player) and the performance/text (virtual world). Customization
and personalization break the *fourth wall*, a term that is borrowed from
dramatic theory. It considers the theatrical stage as having three walls (two
sides and a rear) and an invisible fourth wall boundary between the actors
and audience. Breaking the fourth wall means eliminating the divide
between creator and audience. Today, it also is used to describe the blurred
"boundaries between the fictional and real world, either drawing something *into* the fictional world from outside, or expelling something *out* of
the fictional into the non-fictional" (Conway 2009). In simple terms, the
fourth wall divides the space between the real material world and con-
trived, virtual worlds. In television and theater, the fourth wall is broken
when a character speaks to the audience directly, for instance, thus signify-
ing the artificial nature of the production and yet at the same time allowing
the audience into the character's space through that corecognition of artifice. The actors are "in on the same joke" as the audience, and through
that shared meaning and acknowledgment of artifice, the fourth wall is
broken down.

Player-generated content is typically created outside of the fictional
world (that is, the player makes the object or content outside the diegesis
of the game), but it is then incorporated into the virtual space—much as
someone may create a chair for a dollhouse, which then becomes part of
the story that a child may make up about that space. The player is not in
a position of either audience or actor in a virtual world but holds a position in between, simultaneously being both actor and audience. The fourth
wall is broken in this space as the actor/audience, virtual/real dichotomies
are destroyed. In the case of virtual worlds, the players know that they are
populating the space with objects and characters and producing the story
with their own creations and performances. As described above, allowing
player-generated sonic content may interfere with the feelings of presence
or realism of the virtual space. Players may choose inappropriate music
that contradicts rather than reinforces actions or events, thus drawing
attention to the artifice. In particular, tying lyrics to content in a literal

way (as discussed in the discussion of musical performance in *Second Life*) or tying sonic elements to action in a "Mickey-Moused"[8] fashion might illuminate the artificial nature of the game's construction.

Returning to Yochai Benkler's (2006, 74) concept that creators of virtual worlds "build tools with which users collaborate to tell a story," the shared authorship of virtual worlds takes place in terms of storytelling and object building, but the overall soundscape is very much created by the "audience" of players. The fourth wall that divides the imaginary virtual world from the real world is broken down by the self-reflexive cocreative practices of the players and by the extension of the sonic virtual world into the real-world space. It is not the believability of the virtual world space that makes these worlds attractive to players: engagement with content creation and performative, social interactions are important components of that experience.

Perhaps most important, as discussed in chapter 2, sound always transcends the fourth wall, and thus sound always extends the virtual space into the real space. If we can accept that the magic circle of the game or the fourth wall between the virtual/real is extended into the player's space through the use of sound and through player-generated content (among other techniques), then combining these two phenomena—player-generated sonic content, in other words—suggests that we are missing (or at least, downplaying) some key elements in our current conceptions of immersion.

Players understand the artifice of the space, and rather than breaking engagement, the involvement of the player-audience in the performance-creation act allows the player into that space in ways that contradict the normal player-character divide. Nowhere is this breakdown of the fourth wall more apparent than in the player-generated sonic content of virtual worlds, including sound effects, voice, and music. Player-generated content need not break immersion because players are not ever completely outside the space and therefore cannot ever be completely inside the space. Rather, they may become immersed in the experience of cocreation and play. Just as the audience participation and fourth-wall breaking of Bertoldt Brecht's theater meant that "not only is artifice no obstacle to entertainment but [it] allows additional levels of engagement to occur" (Pinchbeck 2006, 406), virtual worlds allow us to engage in the game space in new ways that are not afforded by offline games.

Rather than viewing the experience as a break in the intended immersive properties of the game, players select and share musical ideas with other players. An equally compelling if different type of experience occurs with the ability to select and share music: the fourth wall of the experience is broken down by the activity. Steven Conway (2009) describes several cases where the fourth wall in video games is intentionally broken by the designers. The game character addresses its player, for example, or dirt spray is left on the screen, implying that players are witnessing the scene through a camera. In such instances, argues Conway, the wall may be broken, but through such activities and an acknowledgment of the distinction between worlds, the "magic circle" is extended to include the player inside the walls of game space. Conway argues that the fourth wall in games does not delineate the space in a manner that breaks the suspension of disbelief (as it may in theater or film) but allows an increased suspension of disbelief by including the player, expanding the magic circle to incorporate those outside elements of play, and extending play beyond the game.

We often talk of the player's immersion in the game, but rather than view the game strictly as a separate space into which players may become immersed, we may more accurately speak of the player being immersed in the game*play*. The act of play, including content creation, leads to the immersive experience. It may be more useful to focus on distinguishing types of immersion that occur in gameplay, such as immersion in the narrative (what might be referred to as *presence*) and immersion in the experience (what might be referred to as *engagement*). Although players may feel less present in virtual worlds in which they partake in player-generated content sharing, the act of creation within that space may lead to a more engaging experience. And rather than viewing engagement as one step on the way to immersion (see, e.g., Brown and Cairns 2004), engagement should be viewed as a different type of immersive experience that is equally as important and meaningful to the player.

Conclusions

I began this book with a series of questions: In what ways do game players interact with sound? What makes interactive sound different from noninteractive sound? What does it mean to interact with sound? And how does interactive sound change players' association to, involvement with, and experience of games? This book has explored these questions from different perspectives by drawing on areas of practice theory and embodied cognition, and through this exploration, it has become clear that interacting with sound is fundamentally different from listening to sound. Interactive sound encourages new listening practices, new technologies, new creative practices, and new ways of engaging with media.

At its most basic level, interactivity alters the ways in which people listen to sound: we attend to sound in ways that may require us to remember and repeat that sound, for example. Our relationship to sound is changed by our ability to self-produce those sounds, whether through evoking, creating, selecting, or shaping the sound. The embodied interaction with sound that occurs in games differentiates the game experience from that of film because interactive sound can make players physically react or respond in a particular way. Players must therefore be more engaged and involved in the sound because if they fail to listen attentively and respond correctly, their game play will suffer.

The ways of interacting with sound in music-based games, for instance, offer players "new modes of musicality" (Miller 2012, 150). Kiri Miller (2012) notes that players hear music differently after playing *Rock Band* (2007) and *Guitar Hero* (2005). Externally to the game, when these players hear songs, they may mentally play along or consider how the song could be transcribed for the game. A comparable approach to player-generated musical content can be found in games. Players may listen to music and imagine where in a game such a song might fit. Indeed, players commonly discuss in online forums the music that they put in the background of their gameplay—sometimes even before the game has been released, suggesting that players are listening to their music and thinking about its potential use in games.[1] Interactivity thus encourages new ways of listening in which players contribute to the sonic environment through their own selecting, shaping, and creating of sound.

The physical involvement of the listener means that interactive sound also involves more (or different) modalities, adding a level of haptic involvement with the sound. This multimodal interaction among image, sound, and action suggests that we need new ways of understanding sound in interactive media that account for physical interaction. I introduced the concept of kinesonic synchresis to describe the three-way system of added value and emergent meaning that may develop from out of these multimodal interactions, but much empirical research and theoretical exploration needs to be undertaken to understand this phenomenon.

Video games may introduce new ways of listening to, creating, and consuming music that go well beyond the game. New media technologies often bring new means of consumption but also bring shifts in the ways in which we desire to consume media. Music videos, for example, led to a desire to watch music (for some at least), and games may lead to a desire to interact with music. Interactive music formats like MXP4 (which allow the user to select separate instrument tracks within songs) have yet to catch on, but in the future our desire to interact with music could lead to new methods of recording and listening practice.

With video games, interaction with sound is a broad concept that goes beyond the playing of the game into altering the sound for players' own creations. Game sound becomes a form of play in the practices described here—including recontextualizations of game sound in other forms of music, covering, sampling, and using game sounds in nongame contexts for the purposes of personal expression. Players like to play with game sound, and interacting with that sound takes a great many forms. Sonic interactivity can mean taking elements out of the game and reusing them (in new songs or in machinima, for instance) or putting new sonic elements into the game (through voice and music, for example). Interactive media by its nature encourages a desire to engage in these types of cocreative practice and to find ways in which the game—and the experience of game sound—can be made our own. By calling into question notions of authorship through these types of cocreativity, the line between professional artist/creator and consumer/player is disintegrating.

These forms of cocreative interactivity extend the life of games, although they present copyright problems. Some scholars have viewed the ferocity of copyright enforcement as a means to suppress political expression. Rebecca Tushnet (2010, 892), for instance, discusses how the remix historically came from minority groups and argues that, "[Copyright law] is a deeply unhealthy system, guaranteeing that citizens attempting to express themselves and participate in cultural and political dialogue can find themselves unexpectedly threatened or silenced by copyright claims." Players' creative practices are endangered by corporate control that may prevent their ability to comment on and engage with their own cultural products. Concerns over intellectual property are understandable, but they also mean that the game creators are failing to tap into the potential to extend the life and popularity of their games. Sound can be a way for developers to make a connection to their players, and developers should work to find ways to encourage the forms of meta-game play activity described here. Such affordances lengthen the shelf life of games and also support new ways of interacting with the product and with other players. Rather than protecting corporate interests, current and proposed copyright laws may serve to damage those interests as players turn away from games that are "locked down" toward more open games with which they can interact in other ways.

Through cocreative practice, games can become a much larger part of our cultural and artistic practice. A common recent question in the game world is, "Are videogames art?" Film critic Roger Ebert (2010) stoked the fires of this debate by repeatedly declaring that games can never be considered art: "I remain convinced that in principle, video games cannot be art. Perhaps it is foolish of me to say 'never,' because never, as Rick Wakeman informs us, is a long, long time. Let me just say that no video gamer now living will survive long enough to experience the medium as an art form." But what Ebert misses is that there was a shift in art in the late twentieth century from objects to practice. This shift has been referred to as a change of focus from objects to action or doing—a shift to an aesthetics of relationships (Green 2010, 2). In this way, Alfred Gell (1998, 5) redefines art as the "social relations in the vicinity of objects mediating social agency . . . between persons and things, and persons and persons via things." Nicolas Bourriaud (2002, 14) has called this shift relational aesthetics, "an art taking as its theoretical horizon the realm of human interactions and its social context, rather than the assertion of an independent and private symbolic space," which suggests "a radical upheaval of the aesthetic, cultural, and political goals introduced by modern art." Bourriaud's description echoes the quotation from Yochai Benkler (1998, 113) in the previous chapter, who stated that such art sets the environment and provides the tools for collaborative creation and shared activity. Considered in this way, by providing tools of creativity and interaction, cocreative games can be viewed as an important phase in redefining games as participatory, performative art spaces.

Interactivity changes the ways in which we can become engaged with and immersed in our media. Immersion can be viewed as an extension of the self into the virtual, but some authors suggest that this comes at a cost. As Âli Yurtsever and Umut Burcu Tasa (2009, 5) explain, "Beginning from the 1980s, the 'myth of disembodiment' was the new evangelic way to 'escape from our embodied world' to an alternative cyber-reality. . . . a civilization of identities who left their bodies behind." As with the separation of sound from its source into a schizophonia, our culture seems to have a high degree of anxiety over the experience of technologically mediated immersion. Several films in the 1980s and 1990s, for instance, reflected a contemporary anxiety over losing our physical presence and becoming submerged (to the point of being lost) in the virtual world. In *Tron* (1982), a hacker is abducted into the computer; in *The Lawnmower Man* (1992), a gardener becomes trapped in the virtual world; and in *The Matrix* (1999), what viewers see as reality turns out to be a virtual world. Such popular-culture representations of immersion in a virtual world exhibit a fear or loss of the body and simultaneously indulge in a technofetishism that in a sense celebrates that same loss.[2] In many conceptions of virtual spaces, immersion transcends the body, and our "amputated" meat is left behind while we mentally engage with this other world. In Marshall McLuhan's (McLuhan and Powers 1989) terms, the result is a "discarnality" or disembodied being.

Recent thinking about embodiment and technology has proposed that rather than disembodying us, technology offers an extended body. Mark Hansen (2006, 95), for instance, suggests that "Because human embodiment no longer coincides with the boundaries of the human body, a disembodiment of the body forms the condition of possibility for a collective (re)embodiment through technics. The human today is embodied in and through technics." In this way, players' avatars "need not be seen as disembodied virtual entities where we leave the corporeal 'meat' body behind, but rather as complex new expressions of prosthetic re-embodiment through which our physical bodies and subjectivities extend themselves into the virtual. Indeed, with the emergence of the digital avatar, narratives of disembodied subjectivities or consciousnesses roaming through virtual reality and cyberspace have largely been replaced by a renewed interest in the body and an awareness of the importance of embodiment in virtual spaces" (Cleland 2008, 209). We are not disembodied so much as we are reembodied through the game space. We are in no danger of leaving our meat behind. One of the clearest examples of this extension of the body into the virtual world is through sound. As players create, evoke, shape, and voice sound in virtual space, the body and its physicality are brought into the virtual through sound.

Future Directions in Interactive Sound Studies

Video games are more than just another medium of expression, means of constructing worlds or generating stories, or source of material for the imagination. Like other forms of narrative media, video games have generated new collective cultural legends, new icons, and new aesthetics. The importance of sound to this cultural phenomenon is largely overlooked and yet is at the heart of much interactive activity in and around games.

Interactivity creates a fundamentally different experiential relationship to sound, but I have only begun to explore that difference here. My focus has been on interactive sound in video games, but many other products use interactive sound (such as toys and computer interfaces) to which many of these same ideas, theories, and questions apply. Interactive sound is also finding its way into more noninteractive media forms. For instance, the experimental film *Timecode* (2000) by Mike Figgis had four simultaneous screens of action (like the Automatic Mario mod described in chapter 5). Each screen had its own sound track, with the mix adjusted to the most significant scene at any one time. The DVD release of the film gives viewers access to the audio tracks so that they can choose the mix of the film: each track can be soloed or muted, including the score. With multiple versions of the film recorded (and two shared on the DVD), the film and sound can be remixed by the audience. The BBC has been broadcasting interactive radio dramas. *The Dark Horse* (2003), for instance, was interrupted every three minutes by phone and short message service (SMS) (text messaging) votes that decided the action for the following three minutes. Although these types of interactivity with sound incorporate user input and feedback of a sort, their types of control and delays in response or feedback make them different from the physical interactivity that players encounter with video gameplay. But how do these other forms of interactivity relate to what was presented here? Can we apply the same theoretical approaches and ideas about sound in video games to other media?

Much theoretical and empirical research remains to be undertaken to understand our embodied interaction with sound, and there are many more questions than answers that I have provided here. Interactive sound requires new theoretical approaches and terminology that account for the embodied agency that we have over that sound. With sound, we need new ways of exploring the effects of our interactivity on the media. We cannot treat video games as an extension of the cinematic tradition, although games have some cinematic elements. Thus, we cannot rely on language and theory brought from film studies to account for the ways in which players experience sound in games. We must work to forge a new theoretical path to explore and explain the many ways in which we play with sound.

Notes

Preface

1. Casual games require little time commitment or financial investment. Examples include many Web-based games and mobile games.

2. Middleware is a software package that is designed to interface between software applications or between people and an application. Audio middleware generally allows developers to implement sound into their game engine quickly without having to write all of the code from scratch.

Introduction

1. The term *collision detection* refers to the ability of the game engine to detect the intersection of objects in the game.

2. See, for instance, the statistics released by the Entertainment Software Association at http://www.theesa.com.

3. I am simplifying here for brevity. In response to the attacks on computationalist theory, many scholars have begun to incorporate an embodied and interactive perspective into computationalist theory (for an overview, see Scheutz 2002).

Chapter 1

1. With games, the haptic interactions that occur are generally proprioceptic (relating to the position of the body) and kinesthetic (relating to the movement of the body), although the rumble effects in some controllers (rumble pads or rumble packs or even motion chairs) provide tactile (vibratory) feedback. The rumble pad is quite limited beyond intensity of feedback, however. The majority of games use rumble as a subsidiary modality that lets players know when they have crashed, been shot, and so on. A few games employ rumble to provide information that would not otherwise be known. In the *Legend of Zelda* games for the Nintendo 64—*Ocarina of Time* (1998) and *Majora's Mask* (2000)—the rumble pad signals hidden treasure

nearby. Due to the limitations of its use, vibratory haptic feedback is probably less important than the kinesthetic haptic interactions that occur between the player and the sound.

2. The one exception is when audio itself is used as both an input and an output, thereby remaining in a single mode, such as a speech-controlled audio-only interface. Even in this case, it is possible to argue that speech is a different modality than audition.

3. In Cazeau's argument, music does not suffer this same fate because it does not strive to have an overtly representational content, at least in the sense of directly representing narrative.

4. *Kinesonic* (*kinesthetic* + *sonic*) refers to "the physicalization of sound or the mapping of sound to bodily movements" (Wilson-Bokowiec and Bokowiec 2006, 47).

5. However, this is not always the case because an intentional delay may built into the event. These delays may be directly related to the event or indirectly related in that the original event triggers one or more new events before the output.

6. Spelling, grammar, and capitalization have corrected for readability in all Web site citations.

Chapter 2

1. Mark Hansen (2006, 20) makes a similar point in his book *Bodies in Code: Interfaces with New Media,* arguing that extension is a technical mediation of the body schema and going a step further to suggest that the body-in-code is "a body submitted to and constituted by an unavoidable and empowering technical deterritorialization— a body whose embodiment is realized, and can only be realized, in conjunction with technics."

2. The term *acousmatic* refers to sound without a visually apparent source. The term was adopted by film theorist Michel Chion (1994, 73) to refer to sound "in the wings" or off-screen. In this sense, sounds are either directly tied to an image, untied (acousmatic), or move from tied to untied (on to off-screen or vice-versa).

3. Although the Kinect does not use controllers with speakers, a similar effect can be sometimes assumed in the sense that players are creating the sound in their own space. Unlike the button-press controllers and relatively constrained movement of the Xbox's hand controllers (for example), the Kinect requires players to move their entire body so that they make the sound of that movement in their space. Players do not need a speaker to tell them that they just jumped up and down because they make the sound in the real world. This is not always the case—players are not really using a tennis racket or bow and arrow in sports games—but for many movement sounds, players are the "speaker."

4. Direct sounds are sounds that arrive at a listener's ears directly, without any reflections off surfaces, whereas reflected sounds are the reverberations of that sound off objects in space, which creates a short delay and colors the sound through attenuating some of the frequencies.

5. The realism of auditory perspective in games can be limited by the placement of loudspeakers in the room. If I am the character, why are "my" sounds coming from "over there" in the corner of the room? This is a technical limitation of home theaters (which have generally been configured for film), and there are several ways in which this limitation can be overcome. Headphones can bring the sound much closer to "home"—so much so that players may have the inverse problem, in-head localization, in which sounds appear to come from within the listener's head-space rather than externally.

6. There are exceptions, such as *Gran Turismo 5* (2010).

7. Other techniques that are used by games also do this. New stereoscopic 3D techniques create the illusion of the space visually coming out toward the player, and haptic devices may also bring the game out to the player, but sound is the most common and perhaps most important means for drawing the virtual into the material space.

Chapter 3

1. *Guitar Hero III* (2007) was allegedly the first video game to break $1 billion in sales (Thorsen 2009).

2. François Delalande (1988) suggests that there are three types of musical gesture—effective gesture (*geste effecteur*), accompanying gesture (*geste accompagnateur*), and figurative gesture (*geste figuré*) (in Cadoz and Wanderly, 2000, 77–78). Effective gestures are necessary to produce sound, accompanist gestures are not necessarily required to produce the sound but accompany the sound production, and figurative gestures are symbolic gestures of musical imagery in the mind of the performer/listener that are related to previous experience (Iazzetta 2000, 262).

3. Available at http://www.youtube.com/watch?v=veTZBlHraDw.

4. Available at http://www.youtube.com/watch?v=S58gvJF3KoM.

5. Available at http://www.youtube.com/watch?v=veTZBlHraDw.

6. As discussed in the introduction, some of the Mario and Luigi role-playing games use a system whereby a beeping sound is substituted for syllables of text and Mario converses with Luigi in a gibberish Italian.

7. *IGN Boards*, E3 2010: Zelda: Skyward Sword Will Be Orchestrated, http://boards.ign.com/legend_of_zelda/b5188/192942056/p1.

8. *Skyrim Forums*, Should Your Character Have a Voice?, http://skyrimforums.org/threads/should-your-character-have-a-voice-possible-spoiler.130/page-2.

9. *Skyrim Forums*, Should Your Character Have a Voice?, http://skyrimforums.org/threads/should-your-character-have-a-voice-possible-spoiler.130.

10. *Giant Bomb*, *Dragon Age II* Forum: The Dialog Wheel: Why I Hate It with a Passion, http://www.giantbomb.com/dragon-age-ii/61-30995/the-dialogue-wheel-why-i-hate-it-with-a-passion/35-486305.

11. *Skyrim Forums*, Should Your Character Have a Voice?, http://skyrimforums.org/threads/should-your-character-have-a-voice-possible-spoiler.130/page-2.

12. Ibid.

13. *Dragon Age Wiki*, Dialogue Wheel, http://dragonage.wikia.com/wiki/Dialogue_wheel.

14. *BioWare Social Network*, *The Conversation Wheel Is Flawed*, http://social.bioware.com/forum/1/topic/141/index/3128608.

15. Ibid.

16. *Giant Bomb*, *Dragon Age II Forum*, The Dialog Wheel: Why I Hate It with a Passion, http://www.giantbomb.com/dragon-age-ii/61-30995/the-dialogue-wheel-why-i-hate-it-with-a-passion/35-486305.

17. *Mass Effect Dialogue Wheel Generator*, http://portalation.com/funstuff/dialogue.php.

18. *Grinding* refers to the work that sometimes is required to increase a character's experience points, financial situation, and so on. The work that is undertaken on the part of the player in killing off weaker enemies, searching for items, and so on sometimes occurs in a very repetitive and time-consuming manner.

19. In *Unreal Championship* (2002), for instance, the voice channel was designed so that all players were mixed at the same volume, which had the effect of disembodying the voice by eliminating proximity effects (Gibbs, Hew, and Wadley 2004). By spatially placing the voice and leveling the sound according to distance, the "audio mush" becomes much easier to separate into perceptually meaningful sound information. However, Martin R. Gibbs, Kevin Hew, and Greg Wadley (2004, 382) found that a poorly designed proximity algorithm means that a voice may suddenly appear midsentence, "with no sense of a person approaching or receding. Participants found these voices just as disembodied as those in *Unreal Championship*."

20. *Nine Inch Nails Wiki*, http://www.ninwiki.com/I_Am_Trying_To_Believe.

21. See Karen Collins (2012) for more on the links between dystopia and industrial music.

Chapter 4

1. According to the Entertainment Software Association's 2010 statistics, 64 percent of game players play with others in the same physical space and not in online play.

2. ABC Notation is a text-based shorthand music-notation system in which text editors can be used to write music.

3. From the *Mabinogi Data* forums, http://mabidata.net/forum/viewtopic.php?f =20&t=1131. The Stop Online Piracy Act (SOPA) (H.R. 3261) is a bill that was introduced in the U.S. House of Representatives on October 26, 2011. If enacted into law by the U.S. Congress, it would impose severe penalties on anyone who uses copyrighted material.

4. Shoutcast and Icecast are streaming media software systems that allow digital content to be broadcast online.

5. Glaznost, 2001, available at http://vimeo.com/groups/glanzol/videos/24964394.

6. Carmageddon data-bending, *Cementimental*, available at http://www .cementimental.com/carmageddon.html.

7. By Harry Callaghan, known as "Harry101UK."

8. Available at http://www.youtube.com/watch?v=JZIVmKOdrBk.

9. A collection of *Super Mario Bros.* covers is available at http://www.youtube.com/ watch?v=gH6R-DYvTz8&p=1330E06037D284B2.

10. Video Games Live Fan Reviews—Ticketmaster, available at http://reviews .ticketmaster.com/7171/976815/video-games-live-reviews/reviews.htm.

11. Full lyrics are available at http://wow.joystiq.com/2007/01/15/wow-songwatch -rapwing-lair-the-best-song-ever-made.

12. Available at http://arcanewhispers.net/songs/ImJustANoob.mp3.

13. Ninbento is available at http://www.u.arizona.edu/~ksimek/about.html.

14. *Mikro Orchestra*, Circuit Bending, available at http://mikroorchestra.com/press/ circuit_bending.pdf.

Chapter 5

1. *EverQuest* forums, Station.com, available at http://forums.station.sony.com/eq/ posts/list.m?topic_id=114012.

2. World of Warcraft Mods: Soundtrack 1.8, available at http://www.warcraft -mods.com/Soundtrack.html.

3. Available at http://www.youtube.com/watch?v=jwxN8sCIOOE.

4. *Score Hero Version 11*, available at http://www.scorehero.com.

5. Rock Band Network, available at http://www.rockband.com/rock-band-network.

6. For a sample of videos, see Kyle Orland (2007).

7. *nOtbOt*, Low standart, available at http://www.lowstandart.net/static.php?page =notbot.

8. *Mickey Mousing* is a term used in film sound to refer to sound that is so closely synchronized to action that it becomes comical.

Conclusions

1. For instance, I queried Google with the phrase "What music do you listen to while playing *Skyrim*" and discovered discussions that existed in October 2011, a month before the game's release. See, for instance, *Xbox Forums*, http://forums .xbox.com/xbox_forums/xbox_360_games/t_z/elder_scrolls_v_skyrim/f/1733/ p/100176/475375.aspx#475375.

2. This is not a new phenomenon. Plato's cave allegory deals with such a fear and questioning of reality, and science fiction often deals with anxiety over the body and virtual identity.

References

Aarseth, E. 1997. *Cybertext: Perspectives on Ergodic Literature.* Baltimore: Johns Hopkins University Press.

Aarseth, E. 2004. Genre Trouble: Narrativism and the Art of Simulation. In *First-Person: New Media as Story, Performance and Game,* ed. N. Wardrip-Fruin and P. Harrigan, 45–56. Cambridge: MIT Press.

Adams, D. 1999. How to Stop Worrying and Learn to Love the Internet. Douglas Adams.com. http://www.douglasadams.com/dna/19990901-00-a.html.

Altman, R. 1992. Sound Space. In *Sound Theory Sound Practice,* ed. R. Altman, 46–64. New York: Routledge.

Amesley, C. 1989. How to Watch Star Trek. *Cultural Studies* 3 (3): 323–339.

Andrews, G. 2007. Dance Dance Revolution: Taking Back Arcade Space. In *Space Time Play,* ed. F. Borries, S. P. Walz, and M. Böttger, 20–21. Basel: Birkhäuser.

Arca, R.-M. 2010. Thwomp Interview: On Video Game Bands and Niches. *Rock Music by Suite 101.* http://www.suite101.com/content/thwomp-interview-on-video-game-bands-and-niches-a266836.

Arnold, S., and M. Langsman. 2009. He's behind You. *Audio Media* 37: 18.

Auslander, P. 2002. Live from Cyberspace or, I Was Sitting at My Computer This Guy Appeared He Thought I Was a Bot. *PAJ a Journal of Performance and Art* 24 (1): 16–21.

Bailenson, J. N., and N. Yee. 2005. Digital Chameleons: Automatic Assimilation of Nonverbal Gestures in Immersive Virtual Environments. *Psychological Science* 16 (10): 814–819.

Ballas, J. 2007. Self-Produced Sound: Tightly Binding Haptics and Audio. In *Haptics and Audio Interaction Design 2007,* ed. I. Oakley and S. Brewster, 1–8. Berlin: Springer-Verlag.

Banks, J., and M. Deuze. 2009. Co-creative Labour. *International Journal of Cultural Studies* 12 (5): 419–431.

Barbieri, F., A. Buonocore, R. Dalla Volta, and M. Gentilucci. 2009. How Symbolic Gestures and Words Interact with Each Other. *Brain and Language* 110: 1–11.

Barker, J. M. 2009. *The Tactile Eye: Touch and the Cinematic Experience.* Berkeley: University of California Press.

Barthes, R. (1991). *Image, Music, Text.* New York: Hill and Wang. (Original work published in 1977)

Bartle, R. A. 1996. Hearts, Clubs, Diamonds, Spades: Players Who Suit MUDs. *Journal of MUD Research* 1 (1). http://www.mud.co.uk/richard/hcds.htm.

Bartle, R. A. 2003. Not Yet You Fools! *Game Girl Advance.* July 28, 2003. http://www.gamegirladvance.com/2003/07/not-yet-you-fools.html.

Bartle, R. A. 2004. *Designing Virtual Worlds.* Berkeley: New Riders.

Bateman, C., and R. Boon. 2006. *Twenty-first Century Game Design.* Hingham, MA: Charles River Media.

BBC News. 2008. Computer Games Drive Social Ties. September 16. http://news.bbc.co.uk/2/hi/technology/7619372.stm.

Behrenshausen, B. G. 2007. Toward a (Kin)aesthetic of Video Gaming: The Case of Dance Dance Revolution. *Games and Culture* 2 (4): 335–354.

Belton, J. 1985. Technology and Aesthetics of Film Sound. In *Film Sound: Theory and Practice,* ed. E. Weis and J. Belton, 63–72. New York: Columbia University Press.

Benkler, Y. 2006. *The Wealth of Networks: How Social Production Transforms Markets and Freedom.* New Haven: Yale University Press.

Benyon, D., K. Höök, and L. Nigay. 2010. Spaces of Interaction. Paper presented at the ACM-BCS Visions of Computer Science Conference, Edinburgh, UK, April 14–16.

Berg, J. 2009. The Contrasting and Conflicting Definitions of *Envelopment.* Paper presented at the 126th AES Convention, Munich, Germany, May 7–10.

Bianchi-Berthouze, N., W. W. Kim, and D. Patel. 2007. Does Body Movement Engage You More in Digital Game Play? And Why? In *Proceedings of ACII 2007* (LNCS 4738), ed. A. Paiva, R. Prada, and R. W. Picard, 102–113. Berlin: Springer.

Blascovich, J., and J. N. Bailenson. 2011. *Infinite Reality: Avatars, Eternal Life, New Worlds, and the Dawn of the Virtual Revolution.* New York: Morrow.

Bonds, S. 2011. Conference Keynote. IEEE International Games Innovation Conference, Orange, CA, November 2.

Bongers, B. 2000. Physical Interfaces in the Electronic Arts: Interaction Theory and Interfacing Techniques for Real-Time Performance. In *Trends in Gestural Control of*

Music, ed. M. Wanderley and M. Battier, 41–70. Paris: Institut de Recherche et Coordination Acoustique Musique, Centre Pompidou.

Borland, J. 2006. Tomorrow's Games, Designed by Players as They Play. *C/net News*. February 2, 2006. http://news.cnet.com/2100-1043_3-6034630.html.

Botvinick, M., and J. Cohen. 1998. Rubber Hands "Feel" Touch That Eyes See. *Nature* 391 (6669): 756.

Bourriaud, N. 2002. *Esthétique relationnelle*. Dijon: Les Presses du réel.

Bridgett, R. 2010. *From the Shadows of Film Sound*. Self-published.

Brown, E., and P. Cairns. 2004. A Grounded Investigation of Game Immersion. In *ACM Conference on Human Factors in Computing Systems, CHI 2004*, 1297–1300. New York: ACM Press.

Bruns, A. 2007. Produsage: Towards a Broader Framework for User-Led Content Creation. Paper presented at the Sixth ACM SIGCHI Conference on Creativity and Cognition, Washington, DC, June 13–15.

Bullerjahn, C., and M. Güldenring. 1994. An Empirical Investigation of Effects of Film Music Using Qualitative Content Analysis. *Psychomusicology* 13: 99–118.

Butler, M. 2003. Taking It Seriously: Intertextuality and Authenticity in Two Covers by the Pet Shop Boys. *Popular Music* 22: 1–19.

Cadoz, C., and M. M. Wanderley. 2000. Gesture-Music. In *Trends in Gestural Control of Music*, ed. M. Wanderley and M. Battier, 71–93. Paris: Institut de Recherche et Coordination Acoustique Musique, Centre Pompidou.

Cage, J. 1988. *Silence: Lectures and Writings*. Middletown, CT: Wesleyan University Press.

Cameron, D., and J. Carroll. 2009. Encoding Liveness: Performance and Real-Time Rendering in Machinima. Paper presented at the Digital Games Research (DiGRA) Conference, London, September 1–4.

Cannon, R. 2003. Introduction to Game Modification. Paper presented at Plaything: The Language of Gameplay, Sydney, October 8–19.

Cannon, R. 2007. Meltdown. In *Videogames and Art*, ed. A. Clarke and G. Mitchell, 8–53. Bristol: Intellect Books.

Caplan, S., D. Williams, and N. Yee. 2009. Problematic Internet Use and Psychosocial Well-Being among MMO Players. *Computers in Human Behavior* 25 (6): 1312–1319.

Cárdenas, M. 2010. Becoming Dragon: A Transversal Technology Study. *CTheory.Net*. April 29, 2010. http://www.ctheory.net/articles.aspx?id=639.

Cardinali, L., C. Brozzoli, and A. Farnè. 2009. Peripersonal Space and Body Schema: Two Labels for the Same Concept? *Brain Topography* 21: 252–260.

Carlsson, A. 2008. Chip Music: Low-Tech Data Music Sharing. In *From Pac-Man to Pop Music Interactive Audio in Games and New Media*, ed. K. Collins, 153–162. Aldershot: Ashgate.

Carr, N. 2002. An Interview with Anders Carlsson AKA GOTO80. *Remix64*. http://www.remix64.com/interview_anders_carlsson_aka_goto80.html.

Cazeau, C. 2005. Phenomenology and Radio Drama. *British Journal of Aesthetics* 45 (2): 157–174.

Chadabe, J. 2004. Electronic Music and Life. *Organised Sound* 9 (2): 3–6.

Chalk, A. 2011. *World of StarCraft* Hit with Copyright Claim. *The Escapist*. January 19, 2011. http://www.escapistmagazine.com/news/view/107080-World-of-StarCraft-Hit-With-Copyright-Claim.

Chan, A. 2004. CPR for the Arcade Culture: A Case History on the Development of the *Dance Dance Revolution* Community. Undergraduate paper, Stanford University. http://www.stanford.edu/group/htgg/cgi-bin/drupal/?q=node/483.

Chion, M. 1994. *Audio-Vision: Sound on Screen*. New York: Columbia University Press.

Clarke, A., and G. Mitchell. 2007. *Videogames and Art*. Bristol, UK: Intellect.

Cleland, K. 2008. *Image Avatars: Self-Other Encounters in a Mediated World*. Ph.D. diss., University of Technology, Sydney.

Cohen, A. J. 2005. How Music Influences the Interpretation of Film and Video: Approaches from Experimental Psychology. In *Selected Reports in Ethnomusicology: Perspectives in Systematic Musicology 12*, ed. R. A. Kendall and R. W. Savage, 15–36. Los Angeles: UCLA Ethnomusicology Publications.

Coleman, B. 2011. Shut Up and Dance: Reflections on Real-Time Synthesis in Machinima Production. http://cms.mit.edu/people/bcoleman/publications/Coleman%20Machinima%20Reader.pdf.

Collins, K. 2002. The Future Is Happening Already: Industrial Music, Dystopia and the Aesthetic of the Machine. Ph.D. diss., University of Liverpool.

Collins, K. 2008. *Game Sound: An Introduction to the History, Theory, and Practice of Video Game Music and Sound Design*. Cambridge: MIT Press.

Collins, K. 2012. *A Bang, a Whimper, and a Beat: Industrial Music and Dystopia*. New York: Mass Media Music Scholars' Press.

Cone, E. 1968. *Musical Form and Musical Performance*. New York: Norton.

Conway, S. 2009. A Circular Wall? Reformulating the Fourth Wall for Video Games. *Gamasutra*. http://www.gamasutra.com/view/feature/4086/a_circular_wall_reformulating_the_.php.

Cook, N. 1998. *Analysing Musical Multimedia*. Oxford: Oxford University Press.

Coppa, F. 2008. Women, *Star Trek,* and the Early Development of Fannish Vidding. *Transformative Works and Cultures* 1. http://journal.transformativeworks.org/index .php/twc/article/view/44.

Corbett, J. 1990. Free, Single and Disengaged. *October* 1 (2): 79–101.

Couldry, N. 2004. Theorising Media as Practice. *Social Semiotics* 14 (2): 115–132.

Coulton, J. 2011. FAQs about the Portal 2 Song. *Jonathan Coulton*, April 29. http:// www.jonathancoulton.com/2011/04/29/faqs-about-the-portal-2-song.

Cover, R. 2006. Audience Inter/active: Interactive Media, Narrative Control and Reconceiving Audience History. *New Media and Society* 8 (1): 139–158.

Cox, A. 2001. The Mimetic Hypothesis and Embodied Musical Meaning. *Musicae Scientiae* 5 (2): 195–212.

Cox, T. J. 2008. Scraping Sounds and Disgusting Noises. *Applied Acoustics* 69: 1195–1204.

Cullen, B. 2010. A Portfolio of Audiovisual Compositions for the "New Media Every-day." Ph.D. diss., Queen's University Belfast.

Cusic, D. 2005. In Defense of Cover Songs. *Popular Music and Society* 28 (2): 171–177.

Damasio, A. R. 2000. *Descartes' Error: Emotion, Reason, and the Human Brain*. New York: Quill.

Darrington, J. 2011. How to Set Up Ventrilo Music Channel. *eHow*. January 20. http://www.ehow.com/how_7822682_set-up-ventrilo-music-channel.html.

Debevec, K., and E. Iyer. 1986. The Influence of Spokespersons in Altering a Product's Gender Image: Implications for Advertising Effectiveness. *Journal of Advertising* 15 (4): 12–20.

De Preester, H., and M. Tsakiris. 2009. Body-Extension versus Body-Incorporation: Is There a Need for a Body-Model? *Phenomenology and the Cognitive Sciences* 8 (3): 307–319.

Descartes, R. 1998. *Meditations and Other Metaphysical Writings*. Suffolk: Penguin Classics. (Original work published in 1641)

Diamond, E. 1996. *Performance and Cultural Politics*. Oxon, UK: Routledge.

Dourish, P. 2001. *Where the Action Is: The Foundations of Embodied Interaction*. Cambridge: MIT Press.

Download 3K. 2006. Talk in Game Character's Voice Now Possible for Ventrilo Users. http://www.download3k.com/Press-Talk-in-Game-Character-s-Voice-Now-Possible -for.html.

Ducheneaut, N., N. Yee, E. Nickell, and R. J. Moore. 2006. Alone Together? Exploring the Social Dynamics of Massively Multiplayer Online Games. In *Proceedings of the ACM Conference on Human Factors in Computing Systems (CHI 2006)*, 407–416. New York: ACM Press.

Ebert, R. 2010. Video Games Can Never Be Art. *Chicago Sun-Times*, April 16. http://blogs.suntimes.com/ebert/2010/04/video_games_can_never_be_art.html.

Ekman, P., W. V. Friesen, and R. W. Levenson. 1990. Voluntary Facial Action Generates Emotion-Specific Autonomic Nervous System Activity. *Psychophysiology* 27 (4): 363–384.

Fay, T. M., S. Selfon, and T. J. Fay. 2004. *Direct X9 Audio Exposed: Interactive Audio Development*. Plano, TX: Wordware.

Ferreiro, L. 2010. Second Life's Thriving Music Scene. *Los Angeles Times*, June 9. http://articles.latimes.com/2010/jun/09/entertainment/la-et-secondlife-concerts-20100609.

Fiske, J. 1992. The Cultural Economy of Fandom. In *The Adoring Audience: Fan Culture and Popular Media*, ed. L. A. Lewis, 30–49. London: Routledge.

Friedberg, A. 1993. *Window Shopping: Cinema and the Postmodern*. Berkeley: University of California Press.

Fritsch, M., and S. Strötgen. 2012. Relatively Live: How to Identify Live Music Performances. *Music and the Moving Image* 5 (1). http://www.jstor.org/discover/10.5406/musimoviimag.5.1.0047?uid=1008&uid=29684&uid=3739696&uid=2129&uid=2134&uid=2&uid=70&uid=4&uid=62&uid=3739256&sid=56289111383.

Gee, J. P. 2004. *What Video Games Have to Teach Us about Learning and Literacy*. Basingstoke, UK: Palgrave Macmillan.

Gell, A. 1998. *Art and Agency: An Anthropological Theory of Art*. Oxford: Oxford University Press.

Ghazala, Q. R. 2004. The Folk Music of Chance Electronics: Circuit-Bending the Modern Coconut. *Leonardo Music Journal* 14: 97–104.

Gibbs, M. R., K. Hew, and G. Wadley. 2004. Social Translucence of the Xbox Live Voice Channel. In *Entertainment Computing*, ed. M. Rauterberg. 377–385. Berlin: Springer-Verlag.

Gibson, W. 2005. God's Little Toys. *Wired* 13. http://www.wired.com/wired/archive/13.07/gibson.html.

Godøy, R. I. 2010. Gestural Affordances of Musical Sound. In *Musical Gestures: Sound, Movement and Meaning*, ed. R. I. Godøy and M. Leman, 103–125. New York: Routledge.

Godøy, R. I., E. Haga, and A. R. Jensenius. 2006. Playing "Air Instruments": Mimicry of Sound-Producing Gestures by Novices and Experts. *Gesture in Human Computer Interaction and Simulation* 3881: 256–267.

Graft, K. 2011 The New Theory of Horror: *Dead Space 2's* Creative Director Speaks. *Gamasutra.* http://www.gamasutra.com/view/feature/6264/the_new_theory_of _horror_dead_.php.

Green, J.-A. 2010. Interactivity and Agency in Real Time Systems. *Soft Borders Conference and Festival Proceedings*, 1–5. Centro Universitário Belas Artes de São Paulo, October 18–21.

Grimshaw, M. 2007. *The Acoustic Ecology of the First-Person Shooter.* Ph.D. diss., University of Waikato.

Haas, E. 2007. Emerging Multimodal Technology: Role in Enhancing Combat and Civilian System Safety. *Professional Safety* 52 (12): 32–38.

Hall, C. 2011. *Papa Sangre* Review. *148 Apps.* http://www.148apps.com/reviews/ papa-sangre-review.

Halloran, J., Y. Rogers, and G. Fitzpatrick. 2003. From Text to Talk: Multiplayer Games and Voice over IP. In *Proceedings of Level Up: First International Digital Games Research Conference*, 130–142, Utrecht, November 4–6.

Hansen, M. B. N. 2006. *Bodies in Code: Interfaces with New Media.* New York: Routledge.

Harris, C. 2004. Mario vs. Donkey Kong: Nintendo Goes Back to a Classic and Amazingly Fun Game Boy Game to Create a New GBA Experience. *IGN.* http://gameboy .ign.com/articles/518/518351p1.html.

Heidegger, M. 1962. *Being and Time.* New York: Harper & Row. (Original work published in 1927)

Heikkilä, V.-M. 2010. Putting the Demoscene in a Context. http://www.pelulamu .net/countercomplex.

Hertz, G. 2009. Methodologies of Reuse in the Media Arts: Exploring Black Boxes, Tactics and Archaeologies. Paper presented at the Digital Arts and Culture Conference, Irvine, CA, December 12–15.

Holisky, A. 2011. *WoW* Subscriber Numbers Down to 10.3 Million Players. *WoW Insider.* http://wow.joystiq.com/2011/11/08/wow-subscriber-numbers-down-to-10-3 -million-players.

Houpt, S. 2004. FrankenArt. *The Globe and Mail*, May 15, R1.

Höysniemi, J. 2006. International Survey on the *Dance Dance Revolution* Game. *Computers in Entertainment* 4 (2): art. 8.

Huiberts, S., and R. van Tol. 2007. *Crazy Frog Racer. Pretty Ugly Game Sound Study.* http://prettyuglygamesoundstudy.com/?s=crazy+frog+racer.

Huizinga, J. 1955. *Homo Ludens: A Study of Play Element in Culture.* Boston: Beacon Press.

Huron, D. 2002. Listening Styles and Listening Strategies. Paper presented at the Society for Music Theory 2002 Conference, Columbus, OH, November 1. http://www.musiccog.ohio-state.edu/Huron/Talks/SMT.2002/handout.html.

Husserl, E. 1962. *Ideas: General Introduction to Pure Phenomenology.* London: Macmillan. (Original work published in 1931)

Iazzetta, F. 2000. Meaning in Musical Gesture. In *Trends in Gestural Control of Music,* ed. M. Wanderley and M. Battier. Paris: Institut de Recherche et Coordination Acoustique Musique, Centre Pompidou.

Ihde, D. 1979. *Technics and Praxis.* Dordrecht: Reidel.

Ihde, D. 1986. *Consequences of Phenomenology.* New York: New York University Press.

Ihde, D. 2002. *Bodies in Technology.* Minneapolis: University of Minnesota Press.

Ito, M. 2011. Machinima in a Fanvid Ecology. *Journal of Visual Culture* 10 (1): 51–54.

Järvinen, A. 2002. Gran Stylissimo: The Audiovisual Elements and Styles in Computer and Video Games. In *Proceedings of Computer Games and Digital Cultures Conference,* ed. F. Mäyrä, 113–128. Tampere: Tampere University Press.

Jenkins, H. 1990. If I Could Speak with Your Sound: Fan Music, Textual Proximity, and Liminal Identification. *Camera Obscura* 8 (2): 148–175.

Jenkins, H. 1992. *Textual Poachers: Television Fans and Participatory Culture.* London: Routledge.

Jenkins, H. 2006a. *Convergence Culture: Where Old and New Media Collide.* Cambridge: MIT Press.

Jenkins, H. 2006b. How to Watch a Fan-Vid. *Confessions of an Aca-Fan: The Official Weblog of Henry Jenkins.* http://www.henryjenkins.org/2006/09/how_to_watch_a_fanvid.html.

Johnson, A. 1996. The First "Official" *Castle Smurfenstein* Home Page. http://www.evl.uic.edu/aej/smurf.html.

Johnson, P. 2006. Hacking Art: Interview with Cory Arcangel. *Fanzine.* http://www.thefanzine.com/articles/art/53/hacking_art-_interview_with_cory_arcangel/1.

Jones, E. 2005. Thingo of the Week: Micromusic. *Mondo Thingo.* http://www.abc.net.au/thingo/txt/s1083340.htm.

Jones, M. 2005. Composing Space: Cinema and Computer Gaming—The Macro-Mise en Scene and Spatial Composition. Paper presented at the Imaginary Worlds Symposium, Sydney, December. http://www.dab.uts.edu.au/research/conferences/imaginary-worlds/composing_space.pdf.

Jones, M. 2007. Vanishing Point: Spatial Composition and the Virtual camera. *Animation: An Interdisciplinary Journal* 2 (3): 225–243.

Jordan, W. 2007. From Rule-Breaking to ROM-Hacking: Theorizing the Computer Game-as-Commodity Situated Play. In *Proceedings of DiGRA* (September): 708–713.

Jørgensen, K. 2008. Left in the Dark: Playing Computer Games with the Sound Turned Off. In *From Pac-Man to Pop Music*, ed. K. Collins, 163–176. Aldershot: Ashgate.

Juul, J. 2006. *Half Real: Video Games between Real Rules and Fictional Worlds*. Cambridge: MIT Press.

Juul, J. 2008. The Magic Circle and the Puzzle Piece. Paper presented at the Philosophy of Computer Games Conference, Oslo, August 13–15.

Kahn, D. 1999. *Noise, Water, Meat: A History of Voice, Sound and Aurality in the Arts*. Cambridge: MIT Press.

Kastbauer, D. 2011. Audio Implementation Greats #10: Made for the Metronome. *Designing Sound*. http://designingsound.org/2011/01/audio-implementation-greats-10-made-for-the-metronome.

Katigbak, R. 2004. Game On. *Montreal Mirror*, October 27. http://www.montrealmirror.com/2004/102104/nightlife2.html.

Kendon, A. 1997. Gesture. *Annual Review of Anthropology* 26: 109–128.

Keysers, C., B. Wicker, V. Gazzola, J. L. Anton, L. Fogassi, and V. Gallese. 2004. A Touching Sight: SII/PV Activation during the Observation and Experience of Touch. *Neuron* 42: 335–346.

Kirn, P. 2006. Cybersonica: Open Source Fijuu Makes Music in 3D, Navigating with a PS2 Controller. *Create Digital Music*. http://createdigitalmusic.com/2006/05/cybersonica-open-source-fijuu-makes-music-in-3d-navigating-with-a-ps2-controller.

Kirn, P. 2009. Where in the Wii Waggle Is Wanted: Two Other Game Music Control Mappings. *Create Digital Music*. http://createdigitalmusic.com/2009/10/wherein-the-wii-waggle-is-wanted-two-other-game-music-control-mappings.

Klemmer, S. R., B. Hartmann, and L. Takayama. 2006. How Bodies Matter: Five Themes for Interaction Design. Paper presented at the Symposium on Designing Interactive Systems (DIS 2006), Pittsburgh, June 26–28.

Kline, S., N. Dyer-Witheford, and G. de Peuter. 2003. *Digital Play: The Interaction of Technology, Culture and Marketing.* Montreal: McGill-Queen's University Press.

Kohler, E., C. Keysers, M. Alessandra, L. F. Umilta, V. Gallese, and G. Rinolatti. 2002. Hearing Sounds, Understanding Actions: Action Representation in Mirror Neurons. *Science* 2 (297): 846–848.

Kohlrausch, A., and S. van de Par. 2005. Audio-Visual Interaction in the Context of Multi-Media Applications. In *Communication Acoustics*, ed. J. Blauert, 109–138. Berlin: Springer.

Kücklich, J. 2005. Precarious Playbour: Modders and the Digital Games Industry. *Fibreculture* 5. http://five.fibreculturejournal.org/fcj-025-precarious-playbour-modders -and-the-digital-games-industry.

Kushner, D. 2002. The Mod Squad. *Popular Science* (August): 69–73.

Lammes, S. 2008. Spatial Regimes of the Digital Playground: Cultural Functions of Spatial Practices in Computer Games. *Space and Culture* 11 (3): 260–272.

Lastowka, G. 2008. User-Generated Content and Virtual Worlds. *Vanderbilt Journal of Entertainment and Technology Law* 10 (4): 893–917.

Laurel, B. 1991. *Computers as Theatre.* Boston: Addison-Wesley.

Lehrman, P. D. 2009. Using Nintendo Wiimote for Music Production. *E-Musician.* http://emusician.com/tutorials/using-nintendo-wiimote-music-production.

Leman, M. 2008. *Embodied Music Cognition and Mediation Technology.* Cambridge: MIT Press.

Lessig, L. 2004. *Free Culture: How Big Media Uses Technology and the Law to Lock Down Culture and Control Creativity.* New York: Penguin.

Lethem, J. 2007. The Ecstasy of Influence: A Plagiarism. *Harper's Magazine.* http://www.harpers.org/archive/2007/02/0081387.

Lindley, S., J. Le Couteur, and N. Bianchi-Berthouze. 2008. Stirring Up Experience through Movement in Game Play: Effects on Engagement and Social Behaviour. Paper presented at the 2008 SIGCHI Conference on Human Factors in Computing Systems, Florence, April 5–10.

Lipscomb, S. D., and S. M. Zehnder. 2004. Immersion in the Virtual Environment: The Effect of a Musical Score on the Video Gaming Experience. *Journal of Physiological Anthropology and Applied Science* 23 (6): 337–343.

Lombard, M., and J. Snyder-Duch. 2001. Interactive Advertising and Presence: A Framework. *Journal of Interactive Advertising* 1 (2). http://www.jiad.org/article13.

Lowood, H. 2008. Found Technology: Players as Innovators in the Making of Machinima. In *Digital Youth, Innovation, and the Unexpected*, ed. T. McPherson, 165–196. Cambridge: MIT Press.

MacInnis, D. J., and C. W. Park. 1991. The Differential Role of Characteristics of Music on High- and Low-Involvement Consumers' Processing of Ads. *Journal of Consumer Research* 18: 161–173.

MacLeod, C. M., and P. A. MacDonald. 2000. Interdimensional Interference in the Stroop Effect: Uncovering the Cognitive and Neural Anatomy of Attention. *Trends in Cognitive Sciences* 4 (10): 383–391.

Manovich, L. 2001. *The Language of New Media*. Cambridge: MIT Press.

Marcus, T. D. 2007. Fostering Creativity in Virtual Worlds: Easing the Restrictiveness of Copyright for User-Created Content. *New York Law School Review* 52: 67–92.

Marks, L. U. 1999. *The Skin of the Film: Intercultural Cinema, Embodiment, and the Senses*. Durham, NC: Duke University Press.

McCarty, C. 2010. The Ladies of *World of Warcraft*. *MMO Reporter*. http://mmoreporter.com/2010/12/02/the-ladies-of-world-of-warcraft.

McGonigal, J. 2008. Why *I Love Bees*: A Case Study in Collective Intelligence Gaming. In *The Ecology of Games: Connecting Youth, Games, and Learning*, ed. K. Salen, 199–228. Cambridge: MIT Press.

McGurk, H., and J. MacDonald. 1976. Hearing Lips and Seeing Voices. *Nature* 246: 746–748.

McLaren, M. 2003. 8-Bit Punk. *Wired*. http://www.wired.com/wired/archive/11.11/mclaren_pr.html.

McLuhan, M., and L. H. Lapham. 1994. *Understanding Media: The Extensions of Man*. Cambridge: MIT Press. (Original work published in 1964)

McLuhan, M., and B. Nevitt. 1972. *Take Today: The Executive as Dropout*. New York: Harcourt, Brace, Jovanovich.

McLuhan, M., and B. R. Powers. 1989. *The Global Village*. Oxford: Oxford University Press.

Mead, A. 2003. Bodily Hearing: Physiological Metaphors and Musical Understanding. *Journal of Music Therapy* 43 (1): 1–19.

Merleau-Ponty, M. 1998. *Phenomenology of Perception*. London: Routledge. (Original work published in 1945)

Messinger, P. R., X. Ge, E. Stroulia, K. Lyons, K. Smirnov, and M. Bone. 2008. On the Relationship between My Avatar and Myself. *Journal of Virtual Worlds Research* 1 (2). http://journals.tdl.org/jvwr/article/viewArticle/352.

Miller, K. 2007. Jacking the Dial: Radio, Race and Place in Grand Theft Auto. *Ethnomusicology* 51 (3): 402–438.

Miller, K. 2009. Schizophonic Performance: Guitar Hero, Rock Band, and Virtual Virtuosity. *Journal of the Society for American Music* 3 (4): 395–429.

Miller, K. 2012. *Playing Along: Music, Video Games and Networked Amateurs*. New York: Oxford University Press.

Milner, R. M. 2009. Working for the Text: Fan Labor and the New Organization. *International Journal of Cultural Studies* 12 (5): 491–508.

Mitchell, G., and A. Clarke. 2007. Videogame Music: Chiptunes Byte Back? *Proceedings of DIGRA* (September): 4–28.

Moore, A., and R. Dockwray. 2010. The Establishment of the Virtual Performance Space in Rock. *Twentieth-Century Music* 5 (2): 219–241.

Moore, A., P. Schmidt, and R. Dockwray. 2011. A Hermeneutics of Spatialization for Recorded Song. *Twentieth-Century Music* 6 (1): 83–114.

Morris, G. (posting as gwEm). 2004. Open Letter to Malcolm McLaren. *Micromusic*. http://micromusic.net/public_letter_gwEm.html.

Morris, S. 2003. WADs, Bots and Mods: Multiplayer: FPS Games as Co-creative Media. Paper presented at the Level Up: Digital Games Research Conference Proceedings, Utrecht, November 4–6.

Morrison, I., and T. Ziemke. 2005. Empathy with Computer Game Characters: A Cognitive Neuroscience Perspective. In *AISB'05: Proceedings of the Joint Symposium on Virtual Social Agents*, 73–79. Hatfield, UK, April 12–15.

Morse, M. 1998. *Virtualities. Television, Media Art and Cyberculture*. Bloomington: Indiana University Press.

Morse, M. 2003. The Poetics of Interactivity. In *Women, Art and Technology*, ed. J. Malloy. Cambridge: MIT Press.

Mugge, R., H. N. J. Schifferstein, and J. P. L. Schoormans. 2010. Product Attachment and Satisfaction: Understanding Consumers' Post-purchase Behavior. *Journal of Consumer Marketing* 27 (3): 271–282.

Murch, W. 2005. Dense Clarity Clear Density. *Transom Review* 5 (1). http://transom.org/?p=6992.

Murphy, S. C. 2004. Live in Your World, Play in Ours: The Spaces of Video Game Identity. *Journal of Visual Culture* 3 (2): 223–238.

Nacke, L. E., M. N. Grimshaw, and C. A. Lindley. 2010. More Than a Feeling: Measurement of Sonic User Experience and Psychophysiology in a First-Person Shooter Game. *Interacting with Computers* 22: 336–343.

Negativland. n.d. Ideas on Fair Use. *Negativland*. http://www.negativland.com/news/?page_id=23.

Newman, J. 2002. In Search of the Videogame Player. *New Media and Society* 4 (3): 405–422.

Newman, J. 2008. *Playing with Videogames*. Oxon, UK: Routledge.

Niedenthal, P. M. 2007. Embodying Emotion. *Science* 316: 1002–1005.

Nintendo. 2011. Iwata asks: Wii Fit. Wii.com. http://us.wii.com/wii-fit/iwata_asks/vol1_page1.jsp.

Nitsche, M. 2008. *Video Game Spaces: Image, Play, and Structure in 3D Game Worlds*. Cambridge: MIT Press.

Oliver, J. 2006. The Game Is Not the Medium . . . or How to Ignore the Shiny Box. Ljudmila.org. http://www.ljudmila.org/~julian/share/text/The-Game-is-not-the -Medium_Oliver-2006.pdf.

Oliver, J., and S. Pickles. 2007. Fijuu2: A Game-Based Audio-Visual Performance and Composition Engine. In *Proceedings of the 2007 Conference on New Interfaces for Musical Expression (NIME07)*, New York, June 6–10.

Orland, K. 2007. Don't Move! Hacked Mario World Levels Play Themselves. *Joystiq*. http://www.joystiq.com/2007/08/22/dont-move-hacked-mario-world-levels-play -themselves.

Papa Sangre. 2011. http://www.papasangre.com/blog.

Paradiso, J. A., J. Heidemann, and T. G. Zimmerman. 2008. Hacking Is Pervasive. *IEEE Pervasive Computing* 7 (3): 13–15.

Pardo, J. S. 2006. On Phonetic Convergence during Conversational Interaction. *Journal of the Acoustical Society of America* 119 (4): 2382–2393.

Parks Associates. 2006. Casual Gaming Market Update. http://www.parksassociates .com/research/industryreports.htm.

Paul, C. 2003. *Digital Art*. London: Thames and Hudson.

Pearce, C., T. Boellstorff, and B. A. Nardi. 2009. *Communities of Play: Emergent Cultures in Multiplayer Games and Virtual Worlds*. Cambridge: MIT Press.

Peters, D. 2010. Enactment in Listening: Intermedial Dance in EGM Sonic Scenarios and the Listening Body. *Performance Research* 15 (3): 81–87.

Phillips-Silver, J., and L. J. Trainor. 2007. Hearing What the Body Feels: Auditory Encoding of Rhythmic Movement. *Cognition* 105: 533–546.

Pichlmair, M., and F. Kayali. 2007. Levels of Sound: On the Principles of Interactivity in Music Video Games. In *Situated Play: Proceedings of DiGRA 2007 Conference*, 424–430. Toyko, September 4–28.

Pinchbeck, D. 2006. A Theatre of Ethics and Interaction? Bertolt Brecht and Learning to Behave in First-Person Shooter Environments. In *Edutainment 2006, Lecture Notes in Computer Science 3942*, ed. Z. Pan, R. Aylett, H. Diener, X. Jin, S. Göbel, and L. Li. 399–408. Berlin: Springer-Verlag.

Pixl Monster. 2011. Online Gaming Stats. http://www.pixlmonster.com/fireball/online-gaming.

Plass-Oude Bos, D. B., B. Reuderink, H. van de Laar, Å. Gürko, C. Mühl, M. Poel, D. Heylen, and A. Nijholt. 2010. Human-Computer Interaction for BCI Games Usability and User Experience. In *Proceedings of the International Conference on Cyberworlds*, 277–281. Singapore, October 20–22.

Postigo, H. 2008. Video Game Appropriation through Modifications: Attitudes Concerning Intellectual Property among Modders and Fans. *Convergence: The International Journal of Research into New Media Technologies* 14 (1): 59–74.

Postigo, H. 2010. Modding to the Big Leagues: Exploring the Space between Modders and the Game Industry. *First Monday* 15 (5). http://firstmonday.org/htbin/cgiwrap/bin/ojs/index.php/fm/article/view/2972.

Ramachandran, V. S. 2009. The Neurons That Shaped Civilization. *TEDIndia*, November. http://www.ted.com/talks/vs_ramachandran_the_neurons_that_shaped_civilization.html.

Rambusch, J. 2006. The Embodied and Situated Nature of Computer Game Play. Paper presented at the Workshop on the Cognitive Science of Games and Game Play, Vancouver, BC, July 26.

Rochat, P. 1995. Early Development of the Ecological Self. In *The Self in Infancy*, ed. P. Rochat, 53–71. Amsterdam: Elsevier.

Rose, F. 2007. Secret Websites, Coded Messages: The New World of Immersive Games. *Wired* 16 (1). http://www.wired.com/entertainment/music/magazine/16-01/ff_args.

Rowland, D. 2005. Game Development Experiences with Spatial Audio. *Future Play 2005*. Lansing, MI.

Rumsey, F. 2002. Spatial Quality Evaluation for Reproduced Sound: Terminology, Meaning, and a Scene-Based Paradigm. *Journal of the Audio Engineering Society* 50 (9): 651–666.

Salen, K., and E. Zimmerman. 2003. *Rules of Play*. Cambridge: MIT Press.

Salen, K. 2002. *Quake! DOOM! Sims!* Transforming Play: Family Albums and Monster Movies. *Walker Art Center*, October 19. http://www.walkerart.org/gallery9/qds.

Sall, A., and R. E. Grinter. 2007. Let's Get Physical! In, out and around the Gaming Circle of Physical Gaming at Home. *Computer Supported Cooperative Work* 15: 199–229.

Salter, A. M. 2009. Once More a Kingly Quest: Fan Games and the Classic Adventure Genre. *Transformative Works and Cultures* 2. http://journal.transformativeworks.org/index.php/twc/article/viewArticle/33.

Salter, C. 2010. *Entangled: Technology and the Transformation of Performance*. Cambridge: MIT Press.

Saltz, D. Z. 1997. The Art of Interaction: Interactivity, Performativity, and Computers. *Journal of Aesthetics and Art Criticism* 55 (2): 117–127.

Sánchez, J., and M. Lumbreras. 1999. Virtual Environment Interaction through 3D Audio by Blind Children. *Journal of Cyberpsychological Behavior* 2 (1): 101–111.

Sanden, P. (Fall 2009). Hearing Glenn Gould's Body: Corporeal Liveness in Recorded Music. *Current Musicology* 88: 7–34.

Sayre, M. 2008. 22.1 Surround Sound. Whoo! *The Game Composer's Blog*. http://gamenotes.org/2008/07/17/221-surround-sound-whoo.

Schafer, R. M. 1969. *The New Soundscape*. Toronto: Berandol.

Scheutz, M. 2002. *Computationalism: New Directions*. Cambridge: MIT Press.

Shaviro, S. 1993. *The Cinematic Body on the Visceral Event of Film viewing*. Minneapolis: University of Minnesota Press.

Shaviro, S. 2010. The Erotic Life of Machines. *Parallax* 8 (4): 21–31.

Sihvonen, T. 2009. Players Unleashed! Modding *The Sims* and the Culture of Gaming. Ph.D. diss., University of Turku, Turku.

Silverstone, R. 2003. Private Reveries and Public Spaces: Some Thoughts on the Relation between Art and Social Science in an Age of Media and Technology. *Proboscis Cultural Snapshots* 4. http://proboscis.org.uk/publications/SNAPSHOTS_prps.pdf.

Simon, B. 2009. Wii Are Out of Control: Bodies, Game Screens and the Production of Gestural Excess. *Loading . . .* 3 (4). http://journals.sfu.ca/loading/index.php/loading/article/viewArticle/65.

Sklens, M. 2005. A Night with the NESkimos Interview: *Nintendo World Report*. http://www.nintendoworldreport.com/interview/2235.

Smith, J. 2004. I Can See Tomorrow in Your Dance: A Study of Dance Dance Revolution and Music Video Games. *Journal of Popular Music Studies* 16 (1): 58–84.

Sobchack, V. 1992. *The Address of the Eye: A Phenomenology of Film Experience*. Princeton: Princeton University Press.

Sobchack, V. 2004. *Carnal Thoughts: Embodiment and Moving Image Culture*. Berkeley: University of California Press.

Solidoro, A. 2008. Narrative and Performance: Reconceptualizating the Relationship in the Videogames Domain. In *Narrative and Fiction: an Interdisciplinary Approach*, ed. D. Robinson, P. Fisher, N. Gilzean, T. Lee, S. J. Robinson, and P. Woodcock, 53–59. Huddersfield: University of Huddersfield.

Solis, G. 2010. I Did It My Way: Rock and the Logic of Covers. *Popular Music and Society* 33 (3): 297–318.

Sotamaa, O. 2004. Playing It My Way? Mapping The Modder Agency. Paper presented at the Internet Research Conference 5.0, Sussex, UK, September 19–22. http://www.uta.fi/~olli.sotamaa/documents/sotamaa_modder_agency.pdf.

Sotamaa, O. 2007. On Modder Labour, Commodification of Play, and Mod Competitions. *First Monday* 12 (9). http://www.uic.edu/htbin/cgiwrap/bin/ojs/index.php/fm/issue/view/252/showToc.

Steinkuehler, C. A. 2006. Massively Multiplayer Online Video Gaming as Participation in a Discourse. *Mind, Culture, and Activity* 13 (1): 38–52.

Stern, E. 2011. Massively Multiplayer Machinima Mikusuto. *Journal of Visual Culture* 10 (1): 42–60.

Stevens, R., and D. Raybould. 2011. *The Game Audio Tutorial: A Practical Guide to Sound and Music for Interactive Games*. London: Focal Press.

Stockburger, A. 2003. The Game Environment from an Auditive Perspective. Paper presented at the Level Up, Digital Games Research Conference, Utrecht, November 4–6. http://www.audiogames.net/pics/upload/gameenvironment.htm.

Stöcker, C., and J. Hoffmann. 2004. The Ideomotor Principle and Motor Sequence Acquisition: Tone Effects Facilitate Movement Chunking. *Psychological Research* 68 (2): 125–137.

Stroop, J. R. 1935. Studies of Interference in Serial Verbal Reactions. *Journal of Experimental Psychology* 18: 643–662.

Tagg, P. 2000. *Kojak: Fifty Seconds of Television Music—Toward the Analaysis of Affect in Popular Music*. New York: Mass Media Music Scholars' Press.

Tanni, V. 2001. Shooter. *Casz Uidas: Moving Images in Public Space*. http://www.caszuidas.nl/site/main.php?page=works&id=169.

Taylor, L. N. 2002. Video Games: Perspective, Point of View and Immersion. Master's thesis, University of Florida.

Taylor, T. L. 2006. *Play between Worlds: Exploring Online Game Culture*. Cambridge: MIT Press.

Taylor, T. L., and E. Witkowski. 2010. This Is How We Play It: What a Mega-LAN Can Teach Us about Games. *FDG* (June): 19–21.

Théberge, P. 1997. *Any Sound You Can Imagine: Making Music / Consuming Technology.* Middletown, CT: Wesleyan University Press.

Thomas, G. J. 1941. Experimental Study of the Influence of Vision on Sound Localization. *Journal of Experimental Psychology* 28: 167–177.

Thomas, M. J. 1988. The Game-as-Art Form: Historic Roots and Recent Trends. *Leonardo* 21 (4): 421–423.

Thorsen, T. 2009. *Guitar Hero III* Strums Up $1 Billion. *Gamespot.* http://www.gamespot .com/news/guitar-hero-iii-strums-up-1-billion-6203056.

Todd, N., and P. McAngus. 1995. The Kinematics of Musical Expression. *Journal of the Acoustical Society of America* 97 (3): 1940–1949.

Toffler, A. 1970. *Future Shock: A Study of Mass Bewilderment in the Face of Accelerating Change.* London: Bodley Head.

Tomczak, S. 2008. Authenticity and Emulation: Chiptune in the Early Twenty-first Century. Paper presented at the International Computer Music Conference, Belfast, August 24–29.

Totilo, S. 2008. BioWare Tells Us *Dragon Age* Stuff: Explains Lack of Voice, Presence of Origins, Hints at Dragons and Console Versions. *MTV Geek.* http://multiplayerblog .mtv.com/2008/08/04/bioware-tells-us-dragon-age-stuff-explains-lack-of-voice -presence-of-origins-hints-at-dragons-and-console-versions.

Turtle Beach. 2011. Exclusive Interview with *Dead Space 2*'s Audio Director. January 14. http://www.turtlebeach.com/blog/?p=419.

Tushnet, R. 2010. I Put You There: User-Generated Content and Anticircumvention. *Vanderbilt Journal of Entertainment and Technology Law* 12: 889–946.

Vachon, J.-F. 2009. Avoiding Tedium: Fighting Repetition in Game Audio. Paper presented at the Audio Engineering Society (AES) Thirty-fifth International Conference, London, February 11–13.

Wachowski, E. 2007. *WoW* SongWatch: Rapwing Lair, the Best Song Ever Made. *WOW Insider.* http://wow.joystiq.com/2007/01/15/wow-songwatch-rapwing-lair -the-best-song-ever-made.

Wadley, G., M. Gibbs, and P. Benda. 2007. Speaking in Character: Using Voice-over-IP to Communicate within MMORPGs. In *IE 07 Proceedings of the Fourth Australasian Conference on Interactive Entertainment,* Melbourne. http://dl.acm.org/citation.cfm?id=1367980.

Wallach, J. 2003. The Poetics of Electrosonic Presence: Recorded Music and the Materiality of Sound. *Journal of Popular Music Studies* 15 (1): 34–64.

Webb, J. 2010. *Lord of the Rings* Online Column: Freebird! *MMORPG.com.* January 19. http://www.mmorpg.com/gamelist.cfm/game/45/feature/3936/Freebird-.html.

Webster, A. 2010. *Papa Sangre* Is an Audio Only Game That's Downright Terrifying. *Gamezebo.* http://www.gamezebo.com/games/papa-sangre/review.

Westecott, E. 2009. The Player Character as Performing Object. In *Proceedings of DiGRA 2009*, London, September 1–4.

Westerkamp, H. 2002. Linking Soundscape Composition and Acoustic Ecology. *Organised Sound* 7 (1): 51–56.

Wharton, A., and K. Collins. 2011. Subjective Measures of the Influence of Music Personalization on Video Game Play: A Pilot Study. *Game Studies* 11 (2). http://gamestudies.org/1102/articles/wharton_collins.

Whitehead, G. 1991. Holes in the Head: A Theatre for Radio Operations. *Performing Arts Journal* 13 (3): 85–91.

Wilson-Bokowiec, J., and M. A. Bokowiec. 2006. Kinaesonics: The Intertwining Relationship of Body and Sound. *Contemporary Music Review* 25 (1): 47–57.

Wirman, H. 2009. Fan Productivity and Game Fandom. *Transformative Works and Cultures* 3. http://journal.transformativeworks.org/index.php/twc/article/view Article/145/115.

Wolfshead. 2004. Open Letter to SOE at 2004 *EverQuest* Guild Summit. *Wolfshead Online. MMORPG Design and Commentary.* http://www.wolfsheadonline.com/2004 -open-letter-to-everquest.

Wright, T., E. Boria, and P. Breidenbach. 2002. Creative Player Actions in FPS Online Video Games: Playing *Counter-Strike. Game Studies* 2 (2). http://www.gamestudies .org/0202/wright.

Young, K. 2009. Sackboy's Voice: Full of Eastern Promise. *Sound Spam.* http://soundspam.blogspot.com/2009/02/sackboys-voice-full-of-eastern-promise.html.

Young, K. 2010a. Interactive Audio Crimes in *Heavy Rain. Sound Spam.* http://soundspam.blogspot.com/2010/04/interactive-audio-crimes-in-heavy-rain.html.

Young, K. 2010b. Voice in BioWare's *Dragon Age: Origins. Sound Spam.* http://soundspam.blogspot.com/2010/05/voice-in-biowares-dragon-age-origins.html.

Yurtsever, Â., and U. B. Tasa. 2009. Redefining the Body in Cyberculture: Art's Contribution to a New Understanding of Embodiment. In *The Real and the Virtual*, ed. by D. Riha and A. Maj, 3–12. Inter-disciplinary Press. http://www.inter-disciplinary .net/wp-content/uploads/2010/04/cyber4ever2130410.pdf.

Audiovisual References

All Is Full of Love (Chris Cunningham / Björk 1998)

Arcangel, Cory. *Super Mario Clouds* (2002). http://www.coryarcangel.com/things-i -made/supermarioclouds

Armed Assault (Bohemia 2006)

Asheron's Call 2: Fallen Kings (Turbine, 2002)

Assassin's Creed (Ubisoft 2007)

Automatic Mario (using Queen's "Don't Stop Me Now") (unknown author 2009). http://www.break.com/game-trailers/game/new-super-mario-bros/automatic-mario -queens-dont-stop-me-now

Beck. *Game Boy Variations* (Interscope 2005)

Björk. All Is Full of Love. *Homogenic* (One Little Indian 1999)

Bowie, David. Space Oddity. *Space Oddity* (Philips 1969)

Brain Age: Train Your Brain in Minutes a Day! (Nintendo 2005)

Carmageddon (Stainless 1997)

Carmageddon data-bending, by Cementimental. http://www.cementimental.com/ carmageddon.html

Cash, Johnny. Hurt. *American IV: The Man Comes Around* (American 2002)

Castle Smurfenstein (Andrew Johnson, Preston Nevins 1983)

Castlevania (Konami 1986)

Castle Wolfenstein (Muse software 1981)

Chrono Trigger (Square Enix 1995)

City of Heroes (Cryptic Studios 2004)

Combat (Atari 1977)

Corby and Baily. *Gameboy_ultra_F_UK* (2001). *Reconnoitre.net*. http://www.reconnoitre
.net/gameboy/index.php

Counter-Strike (Vivendi 1999)

Crazy Frog Racer 2 (KOCH Media 2006)

Dance Dance Revolution (Konami 1998)

Dance, Voldo, Dance (Chris Brandt 2002). *YouTube*. http://www.youtube.com/watch
?v=tUsnW6QjUi4

The Dark Horse (BBC 2003)

Dead Space (Electronic Arts 2008)

Dead Space 2 (Electronic Arts 2011)

Deus Ex (Ion Storm 2000)

Donkey Kong (Nintendo 1981)

Donkey Konga (Nintendo 2003)

Doom (id 1993)

Doom 3 (id 2004)

DragonAge (BioWare 2009)

Dragon Age: Origins (BioWare 2009)

Dragon Age II (BioWare 2011)

Duke Nukem (Apogee 1991)

Electroplankton (Nintendo 2005)

EverQuest (Sony 1999)

Fallout 3 (Bethesda Game Studios 2008)

FarmVille (Zynga 2009)

Final Fantasy (Square Enix 1997)

gLanzoL by Albert Bertolín. http://nofun-games.com

Fabelmod by Glaznost. http://vimeo.com/24964394

Glitch (Tiny Speck 2011)

Gran Turismo (Polyphony Digital 1997)

Gran Turismo 5 (Polyphony Digital / Sony 2010)

Grand Theft Auto (Rockstar 1997)

Grand Theft Auto: San Andreas (Rockstar 2004)

Guitar Hero (Harmonix 2005)

Guitar Hero III (Neversoft 2007)

Habbo Hotel (Sulake 2000)

Half-Life (Valve 1998)

Half-Life 2 (Valve 2004)

Halo: Combat Evolved (Bungie Software 2001)

Halo 2 (Bungie 2004)

Heavy Rain (Quantic Dream 2010)

Hendrix, Jimi. Kiss the Sky. *Kiss the Sky* (Polydor 1984)

Junkie XL. A Little Less Conversation. *Elvis vs. JXL: Little Less Conversation* (RCA 2002)

Kinect Adventures (Microsoft 2010)

King's Quest (Sierra On-Line 1984)

.kkrieger (.theprodukkt 2004)

Langelaar, W. 2007. *nOtbOt*. http://www.lowstandart.net/static.php?page=notbot

The Lawnmower Man (New Line 1992)

Leeroy Jenkins (Ben Schulz 2005)

Legend of Zelda: Majora's Mask (Nintendo 2000)

Legend of Zelda: Ocarina of Time (Nintendo 1998)

Legend of Zelda: Skyward Sword (Nintendo 2011)

Legend of Zelda: Twilight Princess (Nintendo 2006)

Level 70 Elite Tauren Chieftain. I Am Murloc (Blizzard Entertainment 2008)

Little Big Planet (Media Molecule 2008)

Little Big Planet 2 (Media Molecule 2011)

The Lord of the Rings: The Two Towers (Electronic Arts 2002)

The Lord of the Rings Online: Shadows of Angmar (Turbine 2007)

Lyons, Craig. Winter. *Goodbye You* (Grayscale Records 2009)

Mabinogi (devCat 2004)

Mario and Luigi: Bowser's Inside Story (Nintendo 2009)

Mario Paint (Nintendo 1992)

Mass Effect (BioWare 2007)

The Matrix (Warner Bros 1999)

Maxx Sabretooth. You Can Leave Your Hat On. http://slmusiciansdir.blogspot.com/2011/03/maxx-sabretooth.html

MegaMan (Capcom 1987)

Metal Gear Solid (Konami 1998)

Mike Tyson's Punch Out! (Nintendo 1987)

Minecraft (Mojang 2009)

Mon Oncle (Gaumont 1958)

Mushroom Men (Red Fly 2008)

MySims (EA 2007)

Need for Speed: Shift 2 Unleashed (Electronic Arts 2011)

Nelly. Hot in Here. *Nellyville* (Universal 2002)

New Super Mario Bros. (Nintendo 2006)

New Super Mario Bros. Wii (Nintendo 2009)

Nine Inch Nails. Hurt. *The Downward Spiral* (Nothing 1994)

Oliver, Julian. *Quilted Thought Organ* (2001–2003). http://www.ljudmila.org/~julian/qthoth

Oliver, Julian, and Steven Pickles. *q3apd*. http://julianoliver.com/q3apd

Oliver, Julian, and Steven Pickles. *Fijuu* and *Fijuu 2*. http://www.fijuu.com

Orff, Carl. Carmina Burana. *Carmina Burana* (Grand Gala 1992)

Papa Sangre (Somethin' Else 2010)

Pezouvanis, Elisabeth. Little Mac's Confession. *Nintendo Metal* (self-published 2002). http://www.nintendometal.com/music.htm

Portal (Valve 2007)

Portal 2 (Valve 2011)

Professor Layton (Level-5 2007)

Quake (id 1996)

Quake II (id 1997)

Quake III Arena (id 1999)

Real Sound: Kaze No Regret (Warp 1999)

Red Dead Redemption (Rockstar 2010)

Rock Band (Harmonix 2007)

Second Life (Linden Labs 2003)

Shaunconnery. Rapwing Lair. http://wow.joystiq.com/2007/01/15/wow-songwatch
-rapwing-lair-the-best-song-ever-made

Shore, Howard. *Lord of the Rings* (WEA 2001, 2002, 2003)

Shut Up and Dance (*Sims 2* version) (RLW92 n.d.). http://www.dailymotion.com/
video/x3n279_sims-2-aerosmith-shut-up-and-dance_music

Shut Up and Dance (*Armed Assault* version) (Resonator Productions 2008). *YouTube.*
http://www.youtube.com/watch?v=tUsnW6QjUi4

Silent Hill (Konami 1999)

Simon (Milton Bradley 1978)

The Sims (Maxis 2000)

The Sims 2 (Maxis 2004)

The Sims 3 (Electronic Arts 2009)

SingStar (Sony Computer Entertainment Europe 2004)

Skyrim (Bethesda Game Studios 2011)

Space Oddity (A Sims 2 Music Video)-David Bowie (47 Sims 2006). http://video.google
.com/videoplay?docid=7823024126190739294#

Spore (Electronic Arts 2008)

Starcraft (Blizzard 1998)

Super Madrigal Brothers. *Shakestation* (American Patchwork 2002)

Super Mario Bros. (Nintendo 1985)

Super Mario World (Nintendo 1990)

Tetris (Alexey Pajitnov 1984)

Timecode (Mike Figgis 2000)

Toto. Africa. *Toto IV* (Columbia Records 1982)

Toto's Africa Machinima (Fablesim 2006). http://www.youtube.com/watch?v =NfBUCayjNvI

Tron (Walt Disney Pictures 1982)

Unreal (Epic 1998)

Unreal Championship (Epic 2002)

Unreal Tournament (Epic 1999)

Weezer. Beverly Hills. *Make Believe* (XL 2005)

Weezer Beverly Hills Machinima (unknown 2006). http://www.spike.com/video/ beverly-hills-music/2711306

Wii Fit Plus (Nintendo 2009)

Wii Music (Nintendo 2008)

World of Warcraft (Blizzard 2004)

Zak McKracken and the Alien Mindbenders (Lucasfilm Games 1988)

Index